FINITE ELEMENT PROGRAMS FOR STRUCTURAL VIBRATIONS

C. T. F. Ross

Finite Element Programs for Structural Vibrations

With 50 Figures

Springer-Verlag London Ltd.

C. T. F. Ross, BSc, PhD
School of Systems Engineering, Portsmouth Polytechnic, UK

ISBN 978-1-4471-1888-6   ISBN 978-1-4471-1886-2 (eBook)
DOI 10.1007/978-1-4471-1886-2

Cover illustration: Ch. 8 Fig. 7. Mathematical model of tower.

British Library Cataloguing in Publication Data
Ross, C. T. F. (Carlisle T F), 1935–
  Finite element programs for structural vibrations.
  1. Structures. Analysis. Use of computers
  I. Title
  624.1710285416

Library of Congress Cataloging-in-Publication Data
Ross, C. T. F., 1935–
  Finite element programs for structural vibrations / by C. T. F. Ross.
  p: cm.
Includes bibliographical references and index.

1. Structural dynamics – Computer programs. 2. Finite element method –
Computer programs. I. Title
TA654.R67 1991
624.1'7–dc20   91-14180
                                                                        CIP

© Springer-Verlag London 1991
Originally published by Springer-Verlag London Limited 1991
Softcover reprint of the hardcover 1st edition 1991

Typeset by Keytec Typesetting Ltd, Bridport, Dorset
69/3830-543210 Printed on acid-free paper

To my mother
Mrs Phyllis Helen Ross

Acknowledgements

The author would like to thank his colleagues, and in particular Terry Johns, for their contributions.

His thanks are extended to Professor John Boardman and to Graham White for their support and encouragement.

Last, but not least, the author would like to thank Mrs Lesley Jenkinson for the care and devotion she showed in typing the manuscript.

March 1991   *C. T. F. Ross*

Contents

Notation

Unless otherwise stated, the following notation is used:

A	cross-sectional area
I	second moment of area
I_y	second moment about x–y plane
I_z	second moment about x–z plane
I_P	polar second moment of area of the element's cross-section
J	torsional constant
l, L	length
i, j, k	nodal numbers
E	Young's modulus
G	rigidity modulus
ρ	density
x, y, z	local coordinates
$x^\circ, y^\circ, z^\circ$	global coordinates
u, v, w	displacements in x, y, and z directions respectively
$u^\circ, v^\circ, w^\circ$	displacements in x°, y° and z° directions respectively
n	frequency (Hz)
λ	eigenvalue — $n = \sqrt{(1/\lambda)}/(2\pi)$
$[k]$	elemental stiffness matrix
$[K]$	system stiffness matrix
$[m]$	elemental mass matrix
$[M]$	system mass matrix
$\{\ \}$	a column vector
$i = 1(1)\text{N}$	Cycle "i" from unity in unit increments up to a maximum value of "N", and for each value of "i" read (or print)
x_i, y_i, z_i	the corresponding values of x_i, y_i and z_i

⤷ REPEAT

$2\text{E}11 = 2 \times 10^{11}$
$4.1\text{E}{-}5 = 4.1 \times 10^{-5}$

Conditions for Program Usage

1 The disk may only be used by the book purchaser on his/her microcomputer.

2 A back-up copy of the disk may be made by the book purchaser.

3 The program may not be transferred onto any other machine without the written permission of the author, or in the case of his death, by the written permission of his estate.

4 The software is supplied without any warranties of any kind, whether implied or otherwise.

5 All queries regarding function, performance, or adjustments to the programs should be directed to the author:

> Dr. C. T. F. Ross
> 6 Hurstville Drive
> Waterlooville
> Portsmouth
> Hampshire PO7 7NB
> United Kingdom

> (Telephone: 0705 259304)

Additional Notes

Print-outs of disk content shown in the Appendices of this book have been especially formatted for reproduction; users should not expect their print-outs to match those in this section.

Dimension statements in the disk's source code cannot be adjusted, therefore size of problems undertaken by the user will be limited.

Preface

The vibration of structures is of much interest in many branches of engineering. In general, this is because vibrations cause noise and other undesirable effects and very often lead to fatigue failure. Good structural design usually involves the avoidance of harmful vibrations. However, there are some cases when structural vibrations are used as an aid in engineering; for example, when wet concrete is being prepared, or in the operation of certain transducers. Thus, it can be readily appreciated that many engineers require a good knowledge of structural vibrations.

Today, the complicated sets of simultaneous equations that are necessary in vibration analysis can readily be tackled with the aid of digital computers. Indeed, the fundamental approach to vibration can now be by numerical rather than analytical methods, so the solution can be automatic, apart from the feeding in of the necessary geometrical and material properties of the structure.

This book covers many problems that occur with the vibration of skeletal structures. Six FORTRAN 77 computer programs are presented in Appendices 1–6 and on a disk which is suitable for an IBM PC.

Chap. 1 is on "The Finite Element Method" and Chap. 2 demonstrates, with the aid of hand calculations, the finite element solution of some smaller structures. Chap. 3 is on "The Modular Approach", and Chaps. 4–9 describe the six computer programs, with the aid of a large number of worked examples and figures of the resulting eigenmodes.

The programs are capable of determining the free vibration characteristics of continuous beams and of pin-jointed and rigid-jointed plane and space frames, and grillages. The programs for the continuous beams, plane frames and space trusses have been written in a fairly simplified manner to assist in understanding the method involved. They are suitable for small and medium sized matrices, which cover the vast majority of problems of these types. The program for three-dimensional rigid-jointed frames has been written for medium and large sized problems. It uses a frontal reduction technique, which continuously reduces the sizes of the mass and stiffness matrices.

Chapter 1

The Finite Element Method

This chapter will be concerned with introducing the finite element method, via the method of minimum potential. Elemental stiffness and mass matrices will be derived for rods and beams in both local and global coordinates.

1.1 The Finite Element Method

The finite element method is one of the most powerful methods for solving partial differential equations, particularly when these equations apply over complex shapes with complex boundary conditions.

For example, if a partial differential equation were required to be applied over a complex two-dimensional shape, this shape could be sub-divided into a number of triangles, quadrilaterals and similar shapes, as shown in Fig. 1.1. The process then is to solve this differential equation, via the finite element method, over the triangular or quadrilateral or similar shape, and by considering equilibrium and compatibility, to obtain a number of simultaneous equations, which can be solved with the aid of a computer.

The number of simultaneous equations will depend on the number of nodes and the number of degrees of freedom per node, where the nodes are used to define the shape of each element, and also, the overall shape of the problem in question. In Fig. 1.2 the nodes are shown by the small circles. Elements

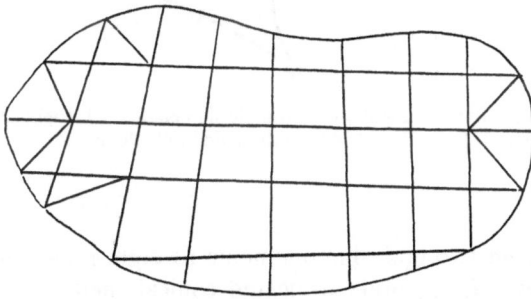

Fig. 1.1. Complex shape, subdivided into triangular and quadrilateral elements.

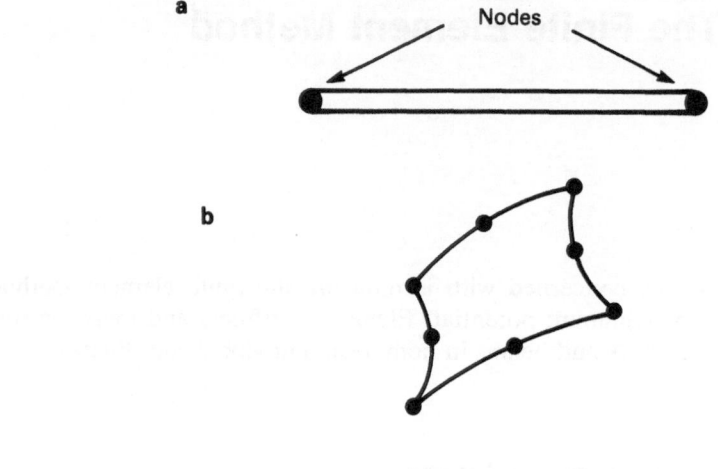

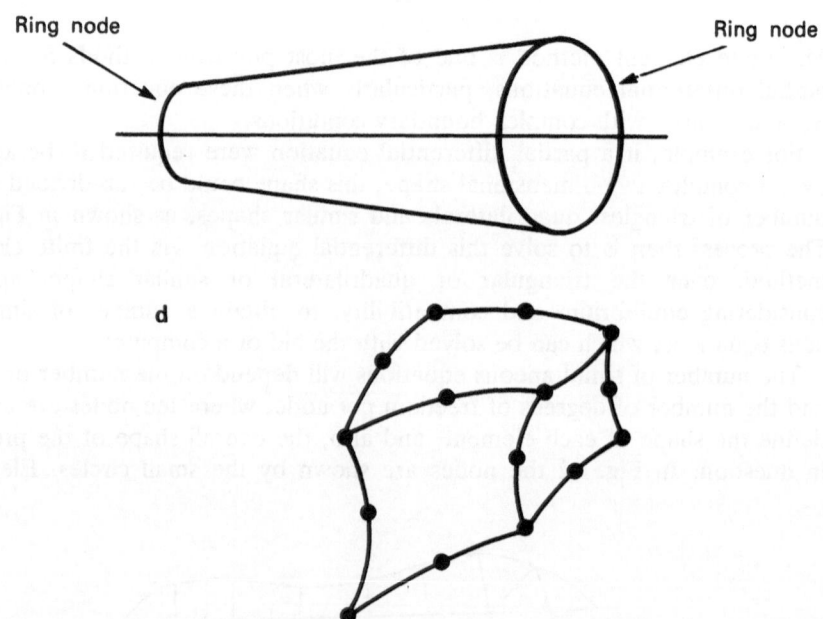

Fig. 1.2. Some typical finite elements. **a** Two-node line element. **b** Eight-node curved quadrilateral. **c** Two-node conical shell element. **d** Twenty-node brick element.

vary in shape from two-node line elements to eight-node curved quadrilaterals, and from two- (ring) node truncated conical shells to twenty-node brick elements, as shown in Fig. 1.2.

The finite element method was first presented by Turner et al. [1], and it was based on the matrix displacement method [2–4], which had previously been used in the aircraft industry. Prior to the presentation of the paper by Turner et al., Courant [5] presented the variational finite difference method, which is another powerful method for solving partial differential equations, particularly in fluid dynamics.

1.2 Derivation of the Elemental Stiffness Matrix of a Rod

If,

$$\pi_p = \text{the total potential}$$

$$= U_e + W$$

then according to the method of minimum potential,

$$\frac{\partial \pi_p}{\partial \{u_i\}} = 0$$

where,

$$U_e = \text{the strain energy of the element}$$

$$W = \text{the potential energy of the load system on the element}$$

$$\{u_i\} = \text{a vector of nodal displacements}$$

Now a rod element is defined as a line element which can only withstand axial loading, so that its strain energy [6] is given by:

$$U_e = \int \frac{\sigma^2}{2E} \, d(\text{vol})$$

$$= \tfrac{1}{2} \int \varepsilon^2 E \, d(\text{vol}) \tag{1.1}$$

where,

$$\sigma = \text{stress}$$

$$\varepsilon = \text{strain}$$

$$E = \text{Young's modulus of elasticity}$$

$$d(\text{vol}) = \text{elemental volume}$$

In matrix form, equation (1.1) appears as

$$U_e = \tfrac{1}{2} \int \{\varepsilon\}^T [D] \{\varepsilon\} \, d(\text{vol}) \tag{1.2}$$

where,

$$\{\varepsilon\} = \text{a vector of strains}$$

$$[D] = \text{a matrix of elastic constants}$$

Now the potential of the external loads is given by:

$$W = -\{u_i\}^{\mathrm{T}}\{P_i\} \tag{1.3}$$

From equations (1.2) and (1.3),

$$\pi_{\mathrm{p}} = \tfrac{1}{2} \int \{\varepsilon\}^{\mathrm{T}}[D]\{\varepsilon\} \, \mathrm{d(vol)} - \{u_i\}^{\mathrm{T}}\{P_i\} \tag{1.4}$$

where,

$$\{P_i\} = \text{a vector of nodal loads}$$

Let,

$$\{\varepsilon\} = [B]\{u_i\} \tag{1.5}$$

where $[B]$ is a matrix that relates the vector of strains, $\{\varepsilon\}$, to the vector of nodal displacements $\{u_i\}$.

Substituting equation (1.5) into equation (1.4):

$$\pi_{\mathrm{P}} = \tfrac{1}{2}\{u_i{}^{\mathrm{T}}\} \int [B]^{\mathrm{T}}[D][B] \, \mathrm{d(vol)}\{u_i\} - \{u_i{}^{\mathrm{T}}\}\{P_i\} \tag{1.6}$$

but the method of minimum potential [7] tells us that

$$\frac{\partial \pi_{\mathrm{p}}}{\partial \{u_i\}} = 0$$

or,

$$\{P_i\} = \int [B^{\mathrm{T}}][D][B] \, \mathrm{d(vol)}\{u_i\} \tag{1.7}$$

Now,

$$\text{Force} = \text{stiffness} \times \text{displacement},$$

or in matrix form,

$$\{P_i\} = [k]\{u_i\} \tag{1.8}$$

where,

$$\{P_i\} = \text{a vector of nodal forces}$$

$$\{u_i\} = \text{a vector of nodal displacements}$$

On comparing equation (1.7) with equation (1.8), it can be seen that

$$[k] = \text{elemental stiffness matrix}$$

$$= \int [B]^{\mathrm{T}}[D][B] \, \mathrm{d(vol)} \tag{1.9}$$

Prior to obtaining $[B]$, it will be necessary to relate the variation of displacement over the element with the vector of nodal displacements.

Consider now the rod element of Fig. 1.3. As the rod element has *two degrees of freedom*, it will be necessary to assume a polynomial for the displacement "u", with *two arbitrary constants*, as follows:

$$u = \alpha_1 + \alpha_2 x \qquad (1.10)$$

where,

α_1 and α_2 are arbitrary constants.

To determine α_1 and α_2, it will be necessary to assume the following *two boundary conditions*:

$$@ \quad x = 0, \quad u = u_1$$
$$@ \quad x = l, \quad u = u_2$$

Substituting the first of these boundary conditions into equation (1.10), gives,

$$\alpha_1 = u_1 \qquad (1.11)$$

Substituting the second of these boundary conditions into equation (1.10):

$$u_2 = \alpha_1 + \alpha_2 l = u_1 + \alpha_2 l$$

or,

$$\alpha_2 = (u_2 - u_1)/l \qquad (1.12)$$

Substituting equations (1.11) and (1.12) into equation (1.10):

$$u = u_1 + (u_2 - u_1)x/l$$
$$= u_1(1 - x/l) + u_2 x/l$$

or,

$$u = u_1(1 - \xi) + u_2\xi \qquad (1.13)$$

where,

$$\xi = x/l = \text{a dimensionless parameter}$$

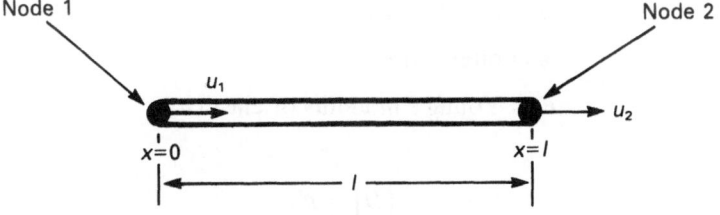

Fig. 1.3.   Two-node rod element.

Equation (1.13) can be put into the matrix form of equation (1.14), i.e.

$$\{u\} = [N]\{u_i\} \tag{1.14}$$

where,

$$\{u\} = \text{a vector of displacement functions}$$

$$[N] = \text{a matrix of shape functions,}$$

which for a *rod element*,

$$[N] = [(1 - \xi) \quad \xi] \tag{1.15}$$

The matrix $[N]$ is called a matrix of shape functions, because it defines the shape of the displacements, and also because it can be used for defining the shape of the element. If the matrix of shape functions for the displacement distribution is the same as the matrix of shape functions for the element shape, the element is said to be *isoparametric*.

From elementary elastic theory [6],

$$\varepsilon = \frac{du}{dx} = \frac{du}{ld\xi}$$

which when applied to equation (1.13), becomes

$$\varepsilon = (-1/l)u_1 + (1/l)u_2 \tag{1.16}$$

In matrix form, equation (1.16) appears as:

$$\{\varepsilon\} = [-1/l \quad 1/l]\{u_i\} \tag{1.17}$$

where for the rod element

$$\{u_i\}^T = [u_1 \quad u_2]$$

Comparing equation (1.17) with equation (1.5):

$$[B] = [-1/l \quad 1/l] \tag{1.18}$$

Similarly, from Hooke's law:

$$\sigma = E\varepsilon$$

where,

$$\sigma = \text{direct stress}$$

$$\varepsilon = \text{direct strain}$$

$$E = \text{Young's modulus of elasticity}$$

i.e.

$$[D] = E \tag{1.19}$$

Substituting equations (1.18) and (1.19) into equation (1.9):

$$[k] = \int_0^1 \begin{bmatrix} -1/l \\ 1/l \end{bmatrix} E[-1/l \quad 1/l] lA\,\mathrm{d}\xi$$

where,

$$A = \text{cross-sectional area of the rod element}$$
$$l = \text{length of the rod element}$$

i.e.

$$[k] = \frac{AEl}{l^2} \begin{bmatrix} 1 & -1 \\ -1 & 1 \end{bmatrix} \int_0^1 \mathrm{d}\xi$$

or,

$$[k] = \frac{AE}{l} \begin{bmatrix} u_1 & u_2 \\ 1 & -1 \\ -1 & 1 \end{bmatrix} \begin{matrix} u_1 \\ u_2 \end{matrix} \tag{1.20}$$

= the elemental stiffness matrix for a rod, where the symbols u_1 and u_2 are used to define the components of the stiffness matrix. N.B. The symbols u_1 and u_2 are not part of $[k]$.

1.3 Derivation of the Elemental Mass Matrix for a Rod

For the dynamic case, it is necessary to introduce the kinetic energy of the element. Now,

$$\text{K.E.} = \text{Kinetic energy} = m\dot{u}^2/2$$

where,

$$m = \text{elemental mass}$$
$$\dot{u} = \text{velocity}$$

or in matrix form:

$$\text{K.E.} = \tfrac{1}{2}\{\dot{u}_i\}^{\mathrm{T}}[m]\{\dot{u}_i\} \tag{1.21}$$

where,

$$[m] = \text{elemental mass matrix}$$
$$\{\dot{u}_i\} = \text{a vector of nodal velocities}$$

Now for the element,

$$\text{K.E.} = \tfrac{1}{2}\int \{\dot{u}\}\rho\{\dot{u}\}\,\mathrm{d}(\mathrm{vol}) \tag{1.22}$$

where,

$\{\dot{u}\}$ = a vector of the velocity variations over the finite element

ρ = density

If SHM takes place

$$\{u\} = Ae^{j\omega t}$$

where,

ω = radian frequency

$j = \sqrt{-1}$

t = time

$\therefore \{\dot{u}\} = j\omega Ae^{j\omega t}$

$= j\omega\{u\}$

but,

$$\{u\} = [N]\{u_i\}$$

$$\therefore \{\dot{u}\} = j\omega[N]\{u_i\} \tag{1.23}$$

which on substituting into equation (1.22), gives

$$\text{K.E.} = -\tfrac{1}{2}\omega^2 \int_{\text{vol}} \{u_i\}^{\text{T}}[N]^{\text{T}}\rho[N]\,\text{d(vol)}\{u_i\} \tag{1.24}$$

Comparing equation (1.21) with equation (1.24)

$[m]$ = the elemental mass matrix

$$= \int [N]^{\text{T}}\rho[N]\,\text{d(vol)} \tag{1.25}$$

Substituting equation (1.15) into equation (1.25),

$[m]$ = the elemental stiffness matrix for a rod

$$= \rho Al \int_0^1 \begin{bmatrix} (1-\xi) \\ \xi \end{bmatrix} [(1-\xi) \quad \xi]\,\text{d}\xi$$

$$= \rho Al \int_0^1 \begin{bmatrix} (1-2\xi+\xi^2) & (\xi-\xi^2) \\ (\xi-\xi^2) & \xi^2 \end{bmatrix}\,\text{d}\xi$$

$$= \rho Al \begin{bmatrix} (1-1+\tfrac{1}{3}) & (\tfrac{1}{2}-\tfrac{1}{3}) \\ (\tfrac{1}{2}-\tfrac{1}{3}) & \tfrac{1}{3} \end{bmatrix}$$

$$[m] = \frac{\rho Al}{6} \begin{matrix} \quad u_1 \quad\; u_2 \\ \begin{bmatrix} 2 & 1 \\ 1 & 2 \end{bmatrix} \begin{matrix} u_1 \\ u_2 \end{matrix} \end{matrix} \tag{1.26}$$

where the displacements u_1 and u_2 are used to define the various components of the mass matrix.

N.B. The symbols u_1 and u_2 are not part of the mass matrix.

1.4 System Stiffness and Mass Matrices

If a structure consists of three in-line rod elements, as shown in Fig. 1.4, where each element has a different cross-sectional area and material property, the system stiffness and mass matrices can be obtained as described below.

From equation (1.20), the elemental stiffness matrices of each element are as follows:

$$[k_{1-2}] = \begin{matrix} u_1 & u_2 \\ \begin{bmatrix} k_1 & -k_1 \\ -k_1 & k_1 \end{bmatrix} & \begin{matrix} u_1 \\ u_2 \end{matrix} \end{matrix} \qquad (1.27)$$

= the elemental stiffness matrix of the element between nodes 1 and 2

$$[k_{2-3}] = \begin{matrix} u_2 & u_3 \\ \begin{bmatrix} k_2 & -k_2 \\ -k_2 & k_2 \end{bmatrix} & \begin{matrix} u_2 \\ u_3 \end{matrix} \end{matrix} \qquad (1.28)$$

= the elemental stiffness matrix of the element between nodes 2 and 3

$$[k_{3-4}] = \begin{matrix} u_3 & u_4 \\ \begin{bmatrix} k_3 & -k_3 \\ -k_3 & k_3 \end{bmatrix} & \begin{matrix} u_3 \\ u_4 \end{matrix} \end{matrix} \qquad (1.29)$$

= the elemental stiffness matrix of the element between nodes 3 and 4

where,

$$k_1 = A_1 E_1 / l_1$$
$$k_2 = A_2 E_2 / l_2$$
$$k_3 = A_3 E_3 / l_3$$

and the suffixes 1, 2 and 3 refer to elements 1–2, 2–3 and 3–4, respectively.

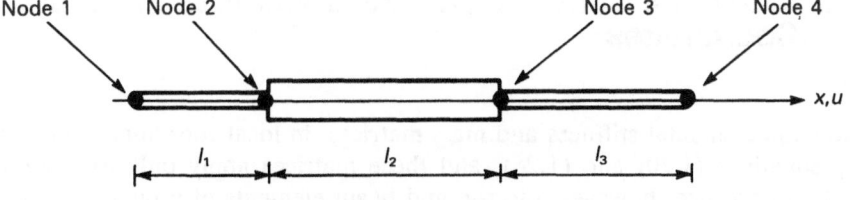

Fig. 1.4.   A three-rod element structure.

The system stiffness matrix $[K]$ can be obtained by forming a matrix of pigeon holes, each pigeon hole corresponding to the displacements u_1, u_2, u_3 and u_4. The next process is to add together the components of stiffness from each elemental stiffness matrix, with respect to the nodal displacements, into the appropriate pigeon hole, as shown by equation (1.30):

$$[K] = \begin{array}{cccc} u_1 & u_2 & u_3 & u_4 \\ \left[\begin{array}{c|c|c|c} k_1 & -k_1 & 0 & 0 \\ \hline -k_1 & k_1 + k_2 & -k_2 & 0 \\ \hline 0 & -k_2 & k_2 + k_3 & -k_3 \\ \hline 0 & 0 & -k_3 & k_4 \end{array}\right] & & & \end{array} \begin{array}{c} u_1 \\ u_2 \\ u_3 \\ u_4 \end{array} \tag{1.30}$$

Similarly, the system mass matrix $[M]$ can be obtained by superimposing the components of each of the three elemental matrices, with respect to the nodal displacements, into the system mass matrix of pigeon holes, as shown by equation (1.31):

$$[M] = \begin{array}{cccc} u_1 & u_2 & u_3 & u_4 \\ \left[\begin{array}{c|c|c|c} m_1 & m_{1/2} & & \\ \hline m_{1/2} & m_1 + m_2 & m_{2/2} & \\ \hline & m_{2/2} & m_2 + m_3 & m_{3/2} \\ \hline & & m_{3/2} & m_4 \end{array}\right] & & & \end{array} \begin{array}{c} u_1 \\ u_2 \\ u_3 \\ u_4 \end{array} \tag{1.31}$$

where,

$$m_1 = \rho_1 A_1 l_1 / 3$$
$$m_2 = \rho_2 A_2 l_2 / 3$$
$$m_3 = \rho_3 A_3 l_3 / 3$$
$$m_4 = \rho_4 A_4 l_4 / 3$$

1.5 Elemental Stiffness and Mass Matrices in Global Coordinates

Now the elemental stiffness and mass matrices, in local coordinates, are given by equations (1.20) and (1.26), and these matrices apply only to horizontal rods. In practice, however, the rod and beam elements of a plane pin-jointed truss or a rigid-jointed plane frame, will normally lie at some angle to the horizontal.

Consider the rod element of Fig. 1.5, which is lying at an angle α to the horizontal, where the $x°$–$y°$ are global reference axes and the x–y axes are the local axes of the rod or beam element.

Consider a point in the positive quadrants of the two systems. From the diagram, it can be readily shown that,

$$u = u° \cos \alpha + v° \sin \alpha$$

and

$$v = -u° \sin \alpha + v° \cos \alpha$$

or in matrix form,

$$\left\{ \begin{array}{c} u \\ v \end{array} \right\} = \left[\begin{array}{cc} c & s \\ -s & c \end{array} \right] \begin{array}{c} u° \\ v° \end{array}$$

or,

$$\{u_i\} = [\zeta]\{u_i°\}, \text{ when applied to the } i\text{th node,}$$

where,

$$c = \cos \alpha$$

$$s = \sin \alpha$$

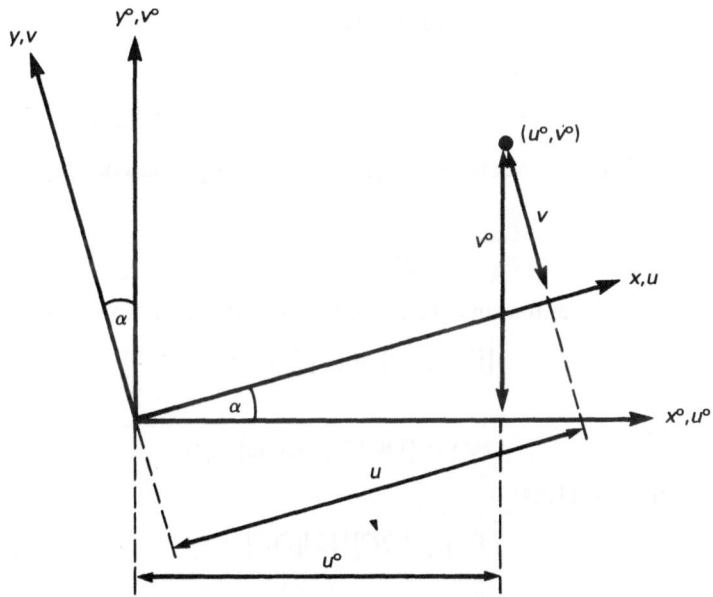

Fig. 1.5.   Global and local coordinate systems.

Now as the rod has two nodal points, the relationship between local and global nodal displacements is given by:

$$\begin{Bmatrix} u_1 \\ v_1 \\ u_2 \\ v_2 \end{Bmatrix} = \left[\begin{array}{c|c} \zeta & O_2 \\ \hline O_2 & \zeta \end{array} \right] \begin{Bmatrix} u_1^\circ \\ v_1^\circ \\ u_2^\circ \\ v_2^\circ \end{Bmatrix} \tag{1.32}$$

or,

$$\{u_i\} = [DC]\{u_i^\circ\} \tag{1.33}$$

where,

$\{u_i\}$ = a vector of nodal displacements in local coordinates
$\{u_i^\circ\}$ = a vector of nodal displacements in global coordinates

$$[DC] = \text{a matrix of directional cosines} = \begin{bmatrix} \zeta & O_2 \\ O_2 & \zeta \end{bmatrix}$$

$$O_2 = \begin{bmatrix} 0 & 0 \\ 0 & 0 \end{bmatrix}$$

Now it can be proven that [DC] is orthogonal [7], so that

$$[DC]^{-1} = [DC]^T \tag{1.34}$$

In a manner similar to that used to determine the expression (1.33), it can be shown that

$$\{P_i\} = [DC]\{P_i^\circ\} \tag{1.35}$$

where,

$\{P_i\}$ = a vector of nodal "forces" in local coordinates

$\{P_i^\circ\}$ = a vector of nodal "forces" in global coordinates

Now,

$$\{P_i\} = [k]\{u_i\} \tag{1.36}$$

and by substituting equations (1.33) and (1.35) into equation (1.36)

$$[DC]\{P_i^\circ\} = [k][DC]\{u_i^\circ\}$$

or,

$$\{P_i^\circ\} = [DC]^{-1}[k][DC]\{u_i^\circ\}$$

but from equation (1.34)

$$\{P_i^\circ\} = [DC]^T[k][DC]\{u_i^\circ\} \tag{1.37}$$

or,

$$\{P_i^\circ\} = [k^\circ]\{u_i^\circ\} \tag{1.38}$$

where,

$$[k^\circ] = [DC]^T[k][DC] \qquad (1.39)$$

Similarly, it can be shown that

$$[m^\circ] = [DC]^T[m][DC] \qquad (1.40)$$

where,

$[k^\circ]$ = elemental stiffness matrix in global coordinates

$[m^\circ]$ = elemental mass matrix in global coordinates

For the two-dimensional rod element of Fig. 1.6:

$$[k] = \frac{AE}{l}\begin{array}{cccc} u_1 & v_1 & u_2 & v_2 \\ \begin{bmatrix} 1 & 0 & -1 & 0 \\ 0 & 0 & 0 & 0 \\ -1 & 0 & 1 & 0 \\ 0 & 0 & 0 & 0 \end{bmatrix} & & & \end{array}\begin{array}{c} u_1 \\ v_1 \\ u_2 \\ v_2 \end{array} \qquad (1.41)$$

and

$$[DC] = \begin{bmatrix} c & s & 0 & 0 \\ -s & c & 0 & 0 \\ 0 & 0 & c & s \\ 0 & 0 & -s & c \end{bmatrix} \qquad (1.42)$$

Substituting equations (1.41) and (1.42) into equation (1.39):

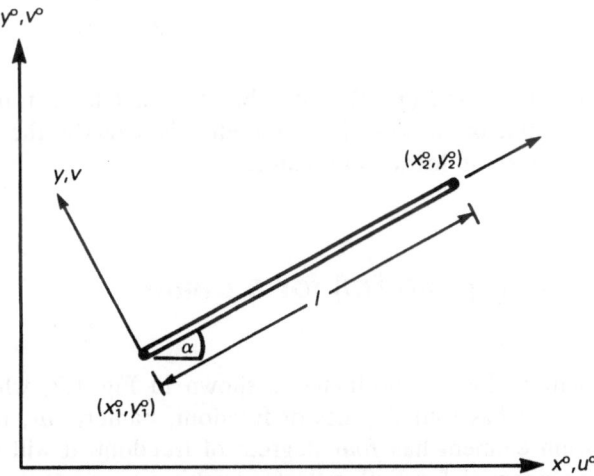

Fig. 1.6.   Two-dimensional rod element.

$$[k^\circ] = \frac{AE}{l} \begin{array}{cccc} u_1^\circ & v_1^\circ & u_2^\circ & v_2^\circ \end{array} \begin{bmatrix} c^2 & cs & -c^2 & -cs \\ cs & s^2 & -cs & -s^2 \\ -c^2 & -cs & c^2 & cs \\ -cs & -s^2 & cs & s^2 \end{bmatrix} \begin{array}{c} u_1^\circ \\ v_1^\circ \\ u_2^\circ \\ v_2^\circ \end{array} \tag{1.43}$$

where,

$$l = \sqrt{[(x_2^\circ - x_1^\circ)^2 + (y_2^\circ - y_1^\circ)^2]} \tag{1.44}$$

$$c = (x_2^\circ - x_1^\circ)/l \tag{1.45}$$

$$s = (y_2^\circ - y_1^\circ)/l \tag{1.46}$$

For a two-dimensional rod element in local coordinates, the elemental mass matrix is given by:

$$[m] = \frac{\rho Al}{6} \begin{array}{cccc} u_1 & v_1 & u_2 & v_2 \end{array} \begin{bmatrix} 2 & 1 & 0 & 0 \\ 1 & 2 & 0 & 0 \\ 0 & 0 & 2 & 1 \\ 0 & 0 & 1 & 2 \end{bmatrix} \begin{array}{c} u_1 \\ v_1 \\ u_2 \\ v_2 \end{array} \tag{1.47}$$

Substituting equations (1.42) and (1.47) into equation (1.40), the elemental mass matrix in *global coordinates* is given by:

$$[m^\circ] = \frac{\rho Al}{6} \begin{array}{cccc} u_1^\circ & v_1^\circ & u_2^\circ & v_2^\circ \end{array} \begin{bmatrix} 2 & 1 & 0 & 0 \\ 1 & 2 & 0 & 0 \\ 0 & 0 & 2 & 1 \\ 0 & 0 & 1 & 2 \end{bmatrix} \begin{array}{c} u_1^\circ \\ v_1^\circ \\ u_2^\circ \\ v_2^\circ \end{array} \tag{1.48}$$

From equations (1.47) and (1.48), it can be seen that for a rod element, the elemental mass matrix in global coordinates is exactly the same as the elemental mass matrix in local coordinates.

1.6 Elemental [k] and [m] for a Beam

A beam element in local coordinates is shown in Fig. 1.7, where it can be seen that the beam has four degrees of freedom, namely, v_1, θ_1, v_2 and θ_2. Now as the beam element has *four* degrees of freedom, it will be convenient to assume a polynomial for "v" with *four* arbitrary constants, as shown by equation (1.49):

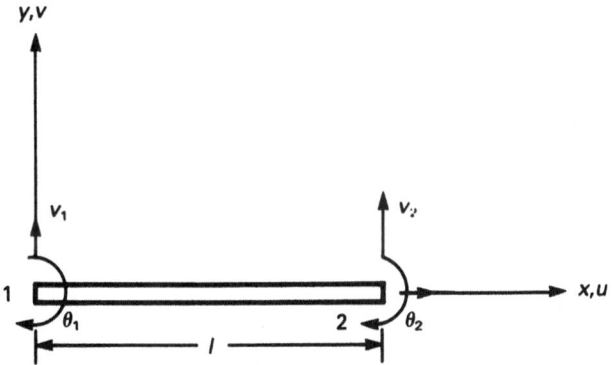

Fig. 1.7. Beam element.

$$v = \alpha_1 + \alpha_2 x + \alpha_3 x^2 + \alpha_4 x^3 \qquad (1.49)$$

Now,

$$\theta = -\frac{dv}{dx} = -\alpha_2 - 2\alpha_3 x - 3\alpha_4 x^2 \qquad (1.50)$$

To determine the *four* arbitrary constants, it will be necessary to obtain *four* simultaneous equations, by assuming the following *four* boundary conditions:

$$@ \ x = 0, \qquad v = v_1$$

$$\therefore \ \alpha_1 = v_1 \qquad (1.51)$$

$$@ \ x = 0, \qquad \theta = \theta_1$$

$$\therefore \ \alpha_2 = -\theta_1 \qquad (1.52)$$

$$@ \ x = l, \qquad v = v_2$$

$$\therefore \ v_2 = v_1 - \theta_1 l + \alpha_3 l^2 + \alpha_4 l^3 \qquad (1.53)$$

$$@ \ x = l, \qquad \theta = \theta_2$$

$$\therefore \ \theta_2 = \theta_1 - 2\alpha_3 l - 3\alpha_4 l^2 \qquad (1.54)$$

Solving equations (1.53) and (1.54):

$$\alpha_3 = 3(v_2 - v_1)/l^2 + (2\theta_1 + \theta_2)/l \qquad (1.55)$$

and

$$\alpha_4 = 2(v_1 - v_2)/l^3 - (\theta_1 + \theta_2)/l^2 \qquad (1.56)$$

Substituting equations (1.51), (1.52), (1.55) and (1.56) into equation (1.49):

$$v = v_1 - \theta_1 x + [3(v_2 - v_1)/l^2 + (2\theta_1 + \theta_2)/l]x^2$$
$$+ [2(v_1 - v_2)/l^3 - (\theta_1 + \theta_2)/l^2]x^3 \qquad (1.57)$$

Rearranging equation (1.57), and putting it in matrix form:

$$v = [(1 - 3\xi^2 + 2\xi^3) \quad l(-\xi + 2\xi^2 - \xi^3)(3\xi^2 - 2\xi^3) \quad l(\xi^2 - \xi^3)][v_1\theta_1v_2\theta_2]^T \tag{1.58}$$

$$= [N]\{u_i\}$$

where,

$$[N] = [(1 - 3\xi^2 + 2\xi^3) \quad l(-\xi + 2\xi^2 - \xi^3)(3\xi^2 - 2\xi^3) \quad l(\xi^2 - \xi^3)] \tag{1.59}$$

$$\xi = x/l$$

It can be shown that equation (1.59) is a Hermite polynomial, osculating to the first degree.

To obtain $[k]$ for a beam element, consider the bending strain energy of the beam:

$$U_b = \text{bending strain energy of the beam}$$

$$= \int \frac{M^2}{2EI} \, dx \tag{1.60}$$

which when compared with (1.2), results in the following analogy:

$$\{\sigma\} = M$$

$$[D] = EI \tag{1.61}$$

$$d(vol) = dx$$

where,

$$M = \text{bending moment}$$

$$E = \text{Young's modulus of elasticity}$$

$$I = \text{second moment of area of the beam's cross-section, about its neutral axis}$$

Furthermore, from [6],

$$EI = \frac{d^2v}{dx^2} = M$$

or,

$$\{\varepsilon\} = \frac{d^2v}{dx^2}$$

In matrix form, equation (1.60) appears as

$$U_b = \tfrac{1}{2} \int \{M\}^T \frac{1}{EI} \{M\} \, dx$$

$$= \tfrac{1}{2} \int \left\{\frac{d^2v}{dx^2}\right\} EI \left\{\frac{d^2v}{dx^2}\right\} dx \tag{1.62}$$

Now,

$$\frac{\mathrm{d}v}{\mathrm{d}x} = \frac{\mathrm{d}v}{l\mathrm{d}\xi} = [(-6\xi + 6\xi^2) \quad l(-1 + 4\xi - 3\xi^2)(6\xi - 6\xi^2) \quad l(2\xi - 3\xi^2)] \begin{Bmatrix} v_1 \\ \theta_1 \\ v_2 \\ \theta_2 \end{Bmatrix}$$

(1.63)

and

$$\frac{\mathrm{d}^2 v}{\mathrm{d}x^2} = \frac{\mathrm{d}^2 v}{l^2 \mathrm{d}\xi^2}$$

$$= [(-6 + 12\xi) \quad l(4 - 6\xi)(6 - 12\xi) \quad l(2 - 6\xi)]\{u_i\} \quad (1.64)$$

$$= [B]\{u_i\}$$

Substituting $[B]$ from equation (1.64) and $[D]$ from equation (1.61) into equation (1.9), and integrating,

$$[k] = EI \begin{matrix} \begin{matrix} v_1 & \theta_1 & v_2 & \theta_2 \end{matrix} \\ \begin{bmatrix} 12/l^3 & -6/l^2 & -12/l^3 & -6/l^2 \\ -6/l^2 & 4/l & 6/l^2 & 2/l \\ -12/l^3 & 6/l^2 & 12/l^3 & 6/l^2 \\ -6/l^2 & 2/l & 6/l^2 & 4/l \end{bmatrix} \begin{matrix} v_1 \\ \theta_1 \\ v_2 \\ \theta_2 \end{matrix} \end{matrix} \quad (1.65)$$

= the elemental stiffness matrix for a beam in local coordinates

To determine $[m]$ for a beam element, substitute $[N]$ from equation (1.59) into equation (1.25) and integrate to give the following:

$$[m] = \frac{\rho Al}{420} \begin{matrix} \begin{matrix} v_1 & \theta_1 & v_2 & \theta_2 \end{matrix} \\ \begin{bmatrix} 156 & & \text{Symmetrical} & \\ -22l & 4l^2 & & \\ 54 & -13l & 156 & \\ 13l & -3l^2 & 22l & 4l^2 \end{bmatrix} \begin{matrix} v_1 \\ \theta_1 \\ v_2 \\ \theta_2 \end{matrix} \end{matrix} \quad (1.66)$$

= elemental mass matrix for a beam in local coordinates

In global coordinates, the elemental stiffness matrix for a beam inclined at an angle α to the x° axis, and in the positive $x^\circ-y^\circ$ quadrant, is given by:

$$[k_b^\circ] = [DC]^T[k][DC]$$

where,

$$[DC] = \begin{bmatrix} \zeta & O_3 \\ \hline O_3 & \zeta \end{bmatrix} \quad (1.67)$$

$$
\zeta =
\begin{array}{cccc}
 & u^{\circ} & v^{\circ} & \theta \\
\end{array}
\left[
\begin{array}{ccc}
c & s & 0 \\
-s & c & 0 \\
0 & 0 & 1 \\
\end{array}
\right]
\begin{array}{c}
u^{\circ} \\
v^{\circ} \\
\theta \\
\end{array}
\tag{1.68}
$$

Substituting equations (1.66) and (1.67) into equation (1.39):

$$
[k_b^{\circ}] = EI
\begin{array}{cccccc}
u_1^{\circ} & v_1^{\circ} & \theta_1 & u_2^{\circ} & v_2^{\circ} & \theta_2 \\
\end{array}
\left[
\begin{array}{cccccc}
\dfrac{12}{l^3}s^2 & & & & & \\[2mm]
-\dfrac{12}{l^3}cs & \dfrac{12}{l^3}c^2 & & \text{Symmetrical} & & \\[2mm]
\dfrac{6}{l^2}s & -\dfrac{6}{l^2}c & \dfrac{4}{l} & & & \\[2mm]
-\dfrac{12}{l^3}s^2 & \dfrac{12}{l^3}cs & -\dfrac{6}{l^2}s & \dfrac{12}{l^3}s^2 & & \\[2mm]
\dfrac{12}{l^3}cs & -\dfrac{12}{l^3}c^2 & \dfrac{6}{l^2}c & -\dfrac{12}{l^3}cs & \dfrac{12}{l^3}c^2 & \\[2mm]
\dfrac{6}{l^2}s & -\dfrac{6}{l}c & \dfrac{2}{l} & -\dfrac{6}{l^2}s & \dfrac{6}{l^2}c & \dfrac{4}{l} \\[2mm]
\end{array}
\right]
\begin{array}{c}
u_1^{\circ} \\
v_1^{\circ} \\
\theta_1 \\
u_2^{\circ} \\
v_2^{\circ} \\
\theta_2 \\
\end{array}
\tag{1.69}
$$

= elemental stiffness matrix for a beam in global coordinates

where,

$$
c = \cos \alpha
$$
$$
s = \sin \alpha
$$

Similarly, the mass matrix for beam element in global coordinates, $[m_b^{\circ}]$, is given by:

$$
[m_b^{\circ}] = [DC]^T [m][DC] \tag{1.70}
$$

$$
[m_b^{\circ}] = \dfrac{\rho A l}{420}
\begin{array}{cccccc}
u_1^{\circ} & v_1^{\circ} & \theta_1 & u_2^{\circ} & v_2^{\circ} & \theta_2 \\
\end{array}
\left[
\begin{array}{cccccc}
156s^2 & & & & & \\
-156cs & 156c^2 & & & & \\
22ls & -22lc & 4l^2 & & & \\
54s^2 & -54cs & 13ls & 156s^2 & & \\
-54cs & 54c^2 & -13lc & -156cs & 156c^2 & \\
-13ls & 13lc & -3l^2 & -22ls & 22lc & 4l^2 \\
\end{array}
\right]
\begin{array}{c}
u_1^{\circ} \\
v_1^{\circ} \\
\theta_1 \\
u_2^{\circ} \\
v_2^{\circ} \\
\theta_2 \\
\end{array}
$$

$$
\tag{1.71}
$$

= elemental mass matrix for a beam in global coordinates

Normally, equations (1.69) and (1.71) cannot be applied directly to an inclined beam element, as the axial effects have to be allowed for in addition to the flexural effects. Thus, when a rigid-jointed plane frame is required to be analysed, it is necessary to superimpose the axial effects of the element, with the aid of the rod element, as follows:

$$[k^\circ] = [k_b^\circ] + [k_r^\circ] \tag{1.72}$$

= elemental stiffness matrix for a beam-column or beam-tie element, inclined at an angle α to the x° axis, as shown in Fig. 1.6.

Similarly,

$$[m^\circ] = [m_b^\circ] + [m_r^\circ] \tag{1.73}$$

= elemental mass matrix for a beam-column or beam-tie element,

where,

$$[k_r^\circ] = \frac{AE}{l} \begin{bmatrix} \overset{u_1^\circ}{c^2} & \overset{v_1^\circ}{cs} & \overset{\theta_1}{0} & \overset{u_2^\circ}{-c^2} & \overset{v_2^\circ}{-cs} & \overset{\theta_2}{0} \\ cs & s^2 & 0 & -cs & -s^2 & 0 \\ 0 & 0 & 0 & 0 & 0 & 0 \\ -c^2 & -cs & 0 & c^2 & cs & 0 \\ -cs & -s^2 & 0 & cs & s^2 & 0 \\ 0 & 0 & 0 & 0 & 0 & 0 \end{bmatrix} \begin{matrix} u_1^\circ \\ v_1^\circ \\ \theta_1 \\ u_2^\circ \\ v_2^\circ \\ \theta_2 \end{matrix}$$

= stiffness matrix for an inclined rod element.

$$[m_r^\circ] = \frac{\rho Al}{6} \begin{bmatrix} \overset{u_1^\circ}{2c^2} & \overset{v_1^\circ}{} & \overset{\theta_1}{} & \overset{u_2^\circ}{} & \overset{v_2^\circ}{} & \overset{\theta_2}{} \\ 2cs & 2s^2 & & \text{Symmetrical} & & \\ 0 & 0 & 0 & & & \\ c^2 & cs & 0 & 2c^2 & & \\ cs & s^2 & 0 & 2cs & 2s^2 & \\ 0 & 0 & 0 & 0 & 0 & 0 \end{bmatrix} \begin{matrix} u_1^\circ \\ v_1^\circ \\ \theta_1 \\ u_2^\circ \\ v_2^\circ \\ \theta_2 \end{matrix}$$

To obtain $[m_r^\circ]$, it was necessary to remove the components of mass in the directions of the "v" displacements as these effects were already included in the derivation for $[m_b^\circ]$ (see [7]).

Chapter 2

Problem Solving in Structural Vibrations

This chapter will be concerned with determining resonant frequencies (eigen-values) and their eigenmodes, for some simple structures. These calculations will reveal the difficulty of solving practical problems without the aid of a computer, which, of course, is the main reason for the presentation of this book.

2.1 Plane Pin-jointed Trusses

Example 2.1

Determine the resonant frequencies and eigenmodes for the plane pin-jointed truss of Fig. 2.1, where the following apply:

$$\rho = 7860 \text{ kg/m}^3 \qquad E = 2 \times 10^{11} \text{ N/m}^2$$

$$A_{1-2} = \text{cross-sectional area of element } 1\text{--}2$$

$$= 4 \times 10^{-4} \text{ m}^2$$

$$A_{2-3} = \text{cross-sectional area of element } 2\text{--}3$$

$$= 4 \times 10^{-4} \text{ m}^2$$

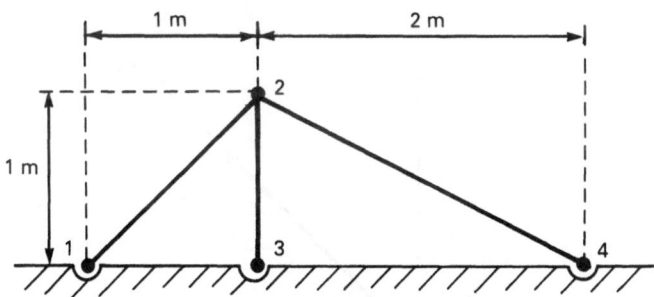

Fig. 2.1. Plane pin-jointed truss.

$$A_{2-4} = \text{cross-sectional area of element 2–4}$$
$$= 6 \times 10^{-4} \text{ m}^2$$

Element 1–2

By defining the element as 1–2, as distinct from 2–1, it means that the element points from 1 to 2, as shown by Fig. 2.2. From equation (1.44),

$$l_{1-2} = \sqrt{[(x_2{}^\circ - x_1{}^\circ)^2 + (y_2{}^\circ - y_1{}^\circ)^2]} = \sqrt{[(1 - 0)^2 + (1 - 0)^2]}$$
$$= 1.414 \text{ m}$$

From equations (1.45) and (1.46),

$$c = \cos \alpha = (x_2{}^\circ - x_1{}^\circ)/l_{1-2} = 1/1.414 = 0.707$$
$$s = \sin \alpha = (y_2{}^\circ - y_1{}^\circ)/l_{1-2} = 1/1.414 = 0.707$$

N.B. If this were defined as 2–1, then the element would point from 2 to 1, so that α would equal 225°.

Substituting the above into equation (1.43) and ignoring the components of stiffness corresponding to the zero displacement, namely, $u_1{}^\circ$ and $v_1{}^\circ$,

$$[k_{1-2}{}^\circ] = \frac{4 \times 10^{-4} \times 2 \times 10^{11}}{1.414} \begin{array}{cccc} u_1{}^\circ & v_1{}^\circ & u_2{}^\circ & v_2{}^\circ \\ & & & \\ \hline & & 0.5 & 0.5 \\ & & 0.5 & 0.5 \end{array} \begin{array}{c} u_1{}^\circ \\ v_1{}^\circ \\ u_2{}^\circ \\ v_2{}^\circ \end{array}$$

$$= 28.29 \times 10^6 \begin{array}{cc} u_2{}^\circ & v_2{}^\circ \\ \begin{bmatrix} 1 & 1 \\ 1 & 1 \end{bmatrix} \end{array} \begin{array}{c} u_2{}^\circ \\ v_2{}^\circ \end{array} \qquad (2.1)$$

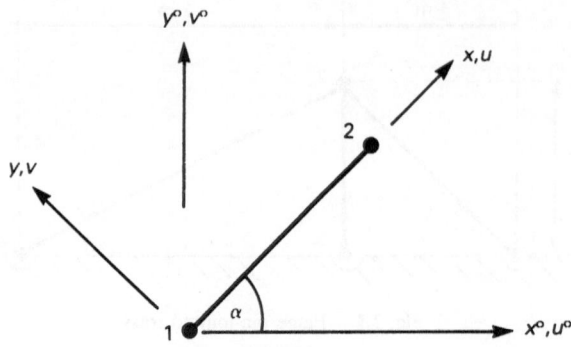

Fig. 2.2. Element 1–2.

Similarly, from equation (1.48),

$$[m_{1-2}{}^\circ] = \frac{7860 \times 4 \times 10^{-4} \times 1.414}{6} \begin{array}{cccc} u_1{}^\circ & v_1{}^\circ & u_2{}^\circ & v_2{}^\circ \end{array} \begin{bmatrix} & & & \\ & & & \\ \hline & & 2 & 0 \\ & & 0 & 2 \end{bmatrix} \begin{array}{l} u_1{}^\circ \\ v_1{}^\circ \\ u_2{}^\circ \\ v_2{}^\circ \end{array}$$

$$= 0.741 \begin{array}{cc} u_2{}^\circ & v_2{}^\circ \end{array} \begin{bmatrix} 2 & 0 \\ 0 & 2 \end{bmatrix} \begin{array}{l} u_2{}^\circ \\ v_2{}^\circ \end{array} \tag{2.2}$$

Element 2–3

The element points from 2 (the start node) to 3 (the finish node), as shown in Fig. 2.3. From equations (1.44)–(1.46),

$$l_{2-3} = \sqrt{[(0 - 0)^2 + (-1 + 0)^2]} = 1 \text{ m}$$
$$c = (x_3{}^\circ - x_2{}^\circ)/l_{2-3} = 0$$
$$s = (y_3{}^\circ - y_2{}^\circ)/l_{2-3} = (-1 - 0)/1 = -1$$

N.B. If this element were defined as 3–2, then the element would point from 3 to 2, so that α would equal 90°.

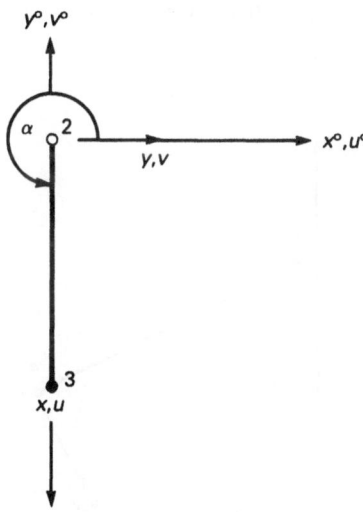

Fig. 2.3. Element 2–3.

Substituting the above into equations (1.43) and (1.48), and ignoring the components of stiffness and mass, corresponding to the zero displacements, namely, u_3° and v_3°,

$$[k_{2-3}^\circ] = 80 \times 10^6 \begin{array}{cccc} u_2^\circ & v_2^\circ & u_3^\circ & v_3^\circ \\ \left[\begin{array}{cc|cc} 0 & 0 & & \\ 0 & 1 & & \\ \hline & & & \\ & & & \end{array}\right] & & & \begin{array}{c} u_2^\circ \\ v_2^\circ \\ \\ u_3^\circ \\ v_3^\circ \end{array} \end{array} \qquad (2.3)$$

$$[m_{2-3}^\circ] = 0.524 \begin{array}{cccc} u_2^\circ & v_2^\circ & u_3^\circ & v_3^\circ \\ \left[\begin{array}{cc|cc} 2 & 0 & & \\ 0 & 2 & & \\ \hline & & & \\ & & & \end{array}\right] & & & \begin{array}{c} u_2^\circ \\ v_2^\circ \\ \\ u_3^\circ \\ v_3^\circ \end{array} \end{array} \qquad (2.4)$$

Element 4–2

The element points from 4 to 2, as shown in Fig. 2.4. From equations (1.44)–(1.46),

$$l_{4-2} = \sqrt{[(3 - 1)^2 + (0 - 1)^2]} = 2.236 \text{ m}$$

$$c = (3 - 1)/2.236 = 0.894$$

$$s = (0 - 1)/2.236 = -0.447$$

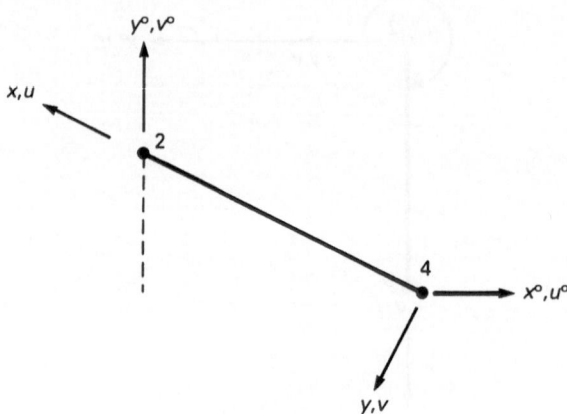

Fig. 2.4.   Element 4–2.

Substituting the above into equations (1.43) and (1.48) and ignoring components of stiffness and mass corresponding to the zero displacements, namely, $u_4{}^\circ$ and $v_4{}^\circ$,

$$[k_{4-2}{}^\circ] = 53.667 \times 10^6 \begin{array}{cccc} u_4{}^\circ & v_4{}^\circ & u_2{}^\circ & v_2{}^\circ \\ \left[\begin{array}{cc|cc} & & & \\ & & & \\ \hline & & 0.8 & -0.4 \\ & & -0.4 & 0.2 \end{array}\right. & & & \left.\begin{array}{c} u_4{}^\circ \\ v_4{}^\circ \\ u_2{}^\circ \\ v_2{}^\circ \end{array}\right] \end{array} \quad (2.5)$$

$$[m_{4-2}{}^\circ] = 1.757 \begin{array}{cccc} u_4{}^\circ & v_4{}^\circ & u_2{}^\circ & v_2{}^\circ \\ \left[\begin{array}{cc|cc} & & & \\ & & & \\ \hline & & 2 & 0 \\ & & 0 & 2 \end{array}\right. & & & \left.\begin{array}{c} u_4{}^\circ \\ v_4{}^\circ \\ u_2{}^\circ \\ v_2{}^\circ \end{array}\right] \end{array} \quad (2.6)$$

Adding together the appropriate components of equations (2.1)–(2.6), the system stiffness and mass matrices are as follows:

$$[K^\circ] = \begin{array}{cc} u_2{}^\circ & v_2{}^\circ \\ \left[\begin{array}{c|c} \begin{array}{c} 28.29 \times 10^6 \\ +0 \\ +42.93 \times 10^6 \end{array} & \begin{array}{c} 28.29 \times 10^6 \\ +0 \\ -21.47 \times 10^6 \end{array} \\ \hline \begin{array}{c} 28.29 \times 10^6 \\ +0 \\ -21.47 \times 10^6 \end{array} & \begin{array}{c} 28.29 \times 10^6 \\ +80 \times 10^6 \\ +10.73 \times 10^6 \end{array} \end{array}\right] & \begin{array}{c} u_2{}^\circ \\ \\ v_2{}^\circ \end{array} \end{array}$$

$$= \begin{bmatrix} 71.22 \times 10^6 & 6.82 \times 10^6 \\ 6.82 \times 10^6 & 119.02 \times 10^6 \end{bmatrix} \quad (2.7)$$

$$[M^\circ] = \begin{array}{cc} u_2{}^\circ & v_2{}^\circ \\ \left[\begin{array}{c|c} \begin{array}{c} 1.482 \\ +1.048 \\ +3.515 \end{array} & \begin{array}{c} 0 \\ +0 \\ +0 \end{array} \\ \hline \begin{array}{c} 0 \\ +0 \\ +0 \end{array} & \begin{array}{c} 1.482 \\ +1.048 \\ +3.515 \end{array} \end{array}\right] & \begin{array}{c} u_2{}^\circ \\ \\ v_2{}^\circ \end{array} \end{array}$$

$$= \begin{bmatrix} 6.045 & 0 \\ 0 & 6.045 \end{bmatrix} \quad (2.8)$$

Substituting equations (2.7) and (2.8) into the dynamic equation:

$$\left| \begin{bmatrix} 71.22 \times 10^6 & 6.82 \times 10^6 \\ 6.82 \times 10^6 & 119.02 \times 10^6 \end{bmatrix} - \omega^2 \begin{bmatrix} 6.045 & 0 \\ 0 & 6.045 \end{bmatrix} \right| = 0$$

which, when expanded becomes

$$(71.22 \times 10^6 - 6.045\omega^2) \times (119.02 \times 10^6 - 6.045\omega^2) - (6.82 \times 10^6)^2 = 0$$

or,

$$8.477 \times 10^{15} - 1.15 \times 10^9 \omega^2 + 36.54\omega^4 - 4.65 \times 10^{13} = 0$$

or,

$$36.54\omega^4 - 1.15 \times 10^9 \omega^2 + 8.43 \times 10^{15} = 0$$

Solving this quadratic equation, yields roots for ω^2, as follows:

$$\omega_1{}^2 = \frac{1.15 \times 10^9 - \sqrt{(9.037 \times 10^{16})}}{73.08}$$

$$= \frac{1.15 \times 10^9 - 3.01 \times 10^8}{73.08} = 11.617 \times 10^6$$

$$\omega_1 = 3409$$

$$n_1 = 542.6 \text{ Hz}$$

$$\omega_2{}^2 = \frac{1.15 \times 10^9 + 3.01 \times 10^8}{73.08} = 19.85 \times 10^6$$

$$\omega_2 = 4455.9$$

$$n_2 = 709.17 \text{ Hz}$$

To Determine the Eigenmodes

Substitute $\omega_1{}^2$ into the first row of the dynamic equation:

$$(71.22 \times 10^6 - 6.045 \times 11.617 \times 10^6)u_2{}^\circ + (6.82 \times 10^6 - 0)v_2{}^\circ = 0$$

or,

$$995235u_2{}^\circ + 6.82 \times 10^6 v_2{}^\circ = 0$$

Let, $u_2{}^\circ = 1$

$$\therefore v_2{}^\circ = -0.146$$

i.e. 1st eigenmode is $[u_2{}^\circ \quad v_2{}^\circ] = [1 \quad -0.146]$

Substitute $\omega_2{}^2$ into the 2nd row of the dynamic equation:

$$(6.82 \times 10^6 - 0)u_2° + (119.02 \times 10^6 - 6.045 \times 19.85 \times 10^6)v_2° = 0$$

$$6.82 \times 10^6 u_2° - 973250v_2° = 0$$

Let, $v_2° = 1$

$$\therefore u_2° = 0.143$$

i.e. 2nd eigenmode is $\quad [u_2° \quad v_2°] = [0.143 \quad 1]$

Example 2.2

Determine the resonant frequencies for the plane pin-jointed truss of Example 2.1, for the case when node 2 has an added mass of 5 kg.

For this case, $[K°]$ is as in equation (2.7) and $[M°]$ is given by:

$$[M°] = \begin{bmatrix} 6.045 & 0 \\ 0 & 6.045 \end{bmatrix} + \begin{bmatrix} 5 & 0 \\ 0 & 5 \end{bmatrix}$$

$$= \begin{matrix} & u_2° & v_2° & \\ & \begin{bmatrix} 11.045 & 0 \\ 0 & 11.045 \end{bmatrix} & \begin{matrix} u_2° \\ v_2° \end{matrix} \end{matrix} \qquad (2.9)$$

Substituting equations (2.7) and (2.9) into the dynamic equation:

$$\left| \begin{bmatrix} 71.22 \times 10^6 & 6.82 \times 10^6 \\ 6.82 \times 10^6 & 119.02 \times 10^6 \end{bmatrix} - \omega^2 \begin{bmatrix} 11.045 & 0 \\ 0 & 11.045 \end{bmatrix} \right| = 0$$

which, when expanded becomes

$$(71.22 \times 10^6 - 11.045\omega^2) \times (119.02 \times 10^6 - 11.045\omega^2) - (6.82 \times 10^6)^2 = 0$$

$$8.477 \times 10^{15} - 2.1 \times 10^9 \omega^2 + 122\omega^4 - 4.65 \times 10^{13} = 0$$

$$122\omega^4 - 2.1 \times 10^9 \omega^2 + 8.431 \times 10^{15} = 0$$

$$\omega_1{}^2 = \frac{2.1 \times 10^9 - \sqrt{(4.41 \times 10^{18} - 4.114 \times 10^{18})}}{244}$$

$$= \frac{2.1 \times 10^9 - 5.438 \times 10^8}{244} = 6.378 \times 10^6$$

$$\omega_1 = 2525$$

$$n_1 = 401.9 \text{ Hz}$$

$$\omega_2{}^2 = \frac{2.1 \times 10^9 + 5.438 \times 10^8}{244} = 10.835 \times 10^6$$

$$\omega_2 = 3291.7$$

$$n_2 = 523.9 \text{ Hz}$$

2.2 Continuous Beams

Example 2.3

Determine the resonant frequencies for the beam of Fig. 2.5, where the following apply:

$$E = 2 \times 10^{11} \text{ N/m}^2 \qquad \rho = 7860 \text{ kg/m}^3$$

$$I_{1-2} = \text{Second moment of area of element } 1\text{–}2$$
$$= 2 \times 10^{-8} \text{ m}^4$$

$$I_{2-3} = \text{Second moment of area of element } 2\text{–}3$$
$$= 3 \times 10^{-8} \text{ m}^4$$

$$A_{1-2} = \text{Cross-sectional area of element } 1\text{–}2$$
$$= 4 \times 10^{-5} \text{ m}^2$$

$$A_{2-3} = \text{Cross-sectional area of element } 2\text{–}3$$
$$= 6 \times 10^{-5} \text{ m}^2$$

Element 1–2

Substituting the appropriate properties into equations (1.65) and (1.66), and ignoring the components of stiffness and mass corresponding to the zero displacements, namely v_1 and θ_1, the elemental stiffness and mass matrices are given by

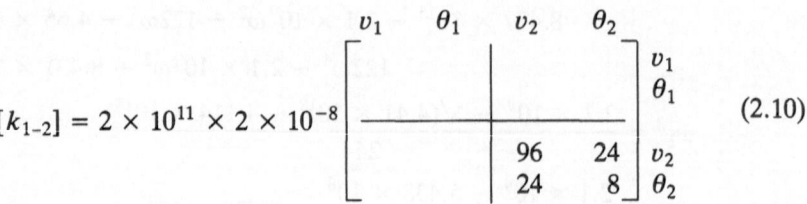

$$[k_{1-2}] = 2 \times 10^{11} \times 2 \times 10^{-8} \qquad (2.10)$$

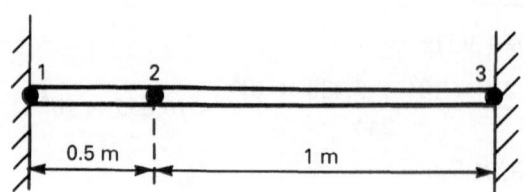

Fig. 2.5. Encastré beam.

$$[m_{1-2}] = \frac{7860 \times 4 \times 10^{-5} \times 0.5}{420} \begin{array}{cccc} v_1 & \theta_1 & v_2 & \theta_2 \\ \left[\begin{array}{cc|cc} & & & \\ \hline & & 156 & 11 \\ & & 11 & 1 \end{array} \right] & \begin{array}{c} v_1 \\ \theta_1 \\ v_2 \\ \theta_2 \end{array} \end{array} \qquad (2.11)$$

Element 2–3

For element 2–3, $l = 1$ m and the elemental stiffness and mass matrices are given by:

$$[k_{2-3}] = \frac{2 \times 10^{11} \times 3 \times 10^{-8}}{1} \begin{array}{cccc} v_2 & \theta_2 & v_3 & \theta_3 \\ \left[\begin{array}{cc|cc} 12 & -6 & & \\ -6 & 4 & & \\ \hline & & & \\ & & & \end{array} \right] & \begin{array}{c} v_2 \\ \theta_2 \\ v_3 \\ \theta_3 \end{array} \end{array} \qquad (2.12)$$

$$[m_{2-3}] = \frac{7860 \times 6 \times 10^{-5} \times 1}{420} \begin{array}{cccc} v_2 & \theta_2 & v_3 & \theta_3 \\ \left[\begin{array}{cc|cc} 156 & -22 & & \\ -22 & 4 & & \\ \hline & & & \\ & & & \end{array} \right] & \begin{array}{c} v_2 \\ \theta_2 \\ v_3 \\ \theta_3 \end{array} \end{array} \qquad (2.13)$$

Superimposing the appropriate coefficients from equations (2.10)–(2.13), the system stiffness matrix $[K^\circ]$ and the system mass matrix $[M^\circ]$ are given by:

$$[K^\circ] = \begin{array}{cc} v_2 & \theta_2 \\ \left[\begin{array}{cc} 0.384 \times 10^6 & 0.96 \times 10^5 \\ +72 \times 10^3 & -36 \times 10^3 \\ & \\ 0.96 \times 10^5 & 0.32 \times 10^5 \\ -36 \times 10^3 & +24 \times 10^3 \end{array} \right] & \begin{array}{c} v_2 \\ \\ \theta_2 \end{array} \end{array}$$

$$= \begin{array}{cc} v_2 & \theta_2 \\ \left[\begin{array}{cc} 0.456 \times 10^6 & 60 \times 10^3 \\ 60 \times 10^3 & 56 \times 10^3 \end{array} \right] & \begin{array}{c} v_2 \\ \theta_2 \end{array} \end{array} \qquad (2.14)$$

$$[M°] = \begin{array}{cc} v_2 & \theta_2 \end{array} \begin{bmatrix} \begin{array}{c} 5.839 \times 10^{-2} \\ +0.175 \end{array} & \begin{array}{c} 4.117 \times 10^{-3} \\ -2.47 \times 10^{-2} \end{array} \\ \begin{array}{c} 4.117 \times 10^{-3} \\ -2.47 \times 10^{-2} \end{array} & \begin{array}{c} 3.743 \times 10^{-4} \\ +4.49 \times 10^{-3} \end{array} \end{bmatrix} \begin{array}{c} v_2 \\ \\ \theta_2 \end{array}$$

$$= \begin{array}{cc} v_2 & \theta_2 \end{array} \begin{bmatrix} 0.233 & -2.058 \times 10^{-2} \\ -2.058 \times 10^{-2} & 4.864 \times 10^{-3} \end{bmatrix} \begin{array}{c} v_2 \\ \theta_2 \end{array} \qquad (2.15)$$

Substituting $[K°]$ and $[M°]$ into the dynamic equation,

$$\begin{bmatrix} (4.56 \times 10^5 - 0.233\omega^2) & (60 \times 10^3 + 2.058 \times 10^{-2}\omega^2) \\ (60 \times 10^3 + 2.058 \times 10^{-2}\omega^2) & (56 \times 10^3 - 4.864 \times 10^{-3}\omega^2) \end{bmatrix} = 0$$

$$(4.56 \times 10^5 - 0.233\omega^2)(56 \times 10^{-3} - 4.864 \times 10^{-3}\omega^2)$$
$$- (60 \times 10^3 + 2.058 \times 10^{-2}\omega^2)^2 = 0$$

$$2.554 \times 10^{10} - 15266\omega^2 + 1.133 \times 10^{-3}\omega^4$$
$$- 3.6 \times 10^9 - 2469.6\omega^2 - 4.235 \times 10^{-4}\omega^4 = 0$$

$$7.095 \times 10^{-4}\omega^4 - 17736\omega^2 + 2.194 \times 10^{10} = 0$$

$$\omega_1{}^2 = \frac{17736 - \sqrt{(3.1457 \times 10^8 - 6.227 \times 10^7)}}{1.419 \times 10^{-3}}$$

$$= \frac{17736 - 15884}{1.419 \times 10^{-3}} = 1.305 \times 10^6$$

$$\omega_1 = 1142$$

$$n_1 = 181.82 \text{ Hz}$$

$$\omega_2{}^2 = \frac{17736 + 15884}{1.419 \times 10^{-3}} = 23.69 \times 10^6$$

$$\omega_2 = 4867.5$$

$$n_2 = 774.7 \text{ Hz}$$

Example 2.4

Determine the magnitudes of the resonant frequencies of the beam of Example 2.3, when node 2 has an additional mass attached to it.

Value of additional mass at node 2 = 1 kg

Value of additional mass moment of inertia at node 2

$$= 0.1 \text{ kg m}^2$$

The system stiffness matrix is the same as equation (2.14), i.e.

$$[K^\circ] = \begin{bmatrix} 4.56\mathrm{E}5 & 60\mathrm{E}3 \\ 60\mathrm{E}3 & 56\mathrm{E}3 \end{bmatrix}$$

The system mass matrix is obtained by adding the components of the concentrated mass to equation (2.15),

$$[M^\circ] = \begin{matrix} & v_2 & \theta_2 \\ & \begin{bmatrix} 0.233 & -2.058\mathrm{E}{-2} \\ -2.058\mathrm{E}{-2} & 4.864\mathrm{E}{-3} \end{bmatrix} \begin{matrix} v_2 \\ \theta_2 \end{matrix} + \begin{matrix} v_2 & \theta_2 \\ \begin{bmatrix} 1 & 0 \\ 0 & 0.1 \end{bmatrix} \end{matrix} \begin{matrix} v_2 \\ \theta_2 \end{matrix} \end{matrix}$$

$$= \begin{bmatrix} 1.233 & -2.058\mathrm{E}{-2} \\ -2.058\mathrm{E}{-2} & 0.105 \end{bmatrix} \tag{2.16}$$

Substituting equations (2.14) and (2.16) into the dynamic equation, the eigenvalue equation becomes:

$$\begin{bmatrix} (4.56 \times 10^5 - 1.233\omega^2) & (60 \times 10^3 + 2.058 \times 10^{-2}\omega^2) \\ (60 \times 10^3 + 2.058 \times 10^{-2}\omega^2) & (56 \times 10^3 - 0.105\omega^2) \end{bmatrix} = 0$$

$$(4.56 \times 10^5 - 1.233\omega^2)(56 \times 10^3 - 0.105\omega^2)$$
$$- (60 \times 10^3 + 2.058 \times 10^{-2}\omega^2)^2 = 0$$

$$2.554 \times 10^{10} - 116\,928\omega^2 + 0.129\omega^4$$
$$- 3.6 \times 10^9 - 2469.6\omega^2 - 4.235 \times 10^{-4}\omega^4 = 0$$

$$0.1286\omega^4 - 119\,398\omega^2 + 2.194 \times 10^{10} = 0$$

$$\omega_1{}^2 = \frac{119\,398 - 54\,497}{0.257} = 252\,532$$

$$\omega_1 = 502.5$$

$$n_1 = 79.98 \text{ Hz}$$

$$\omega_2{}^2 = 676\,634$$

$$\omega_2 = 822.6$$

$$n_2 = 130.9 \text{ Hz}$$

2.3 Pin-jointed Space Trusses

Example 2.5

Determine the resonant frequencies for the pin-jointed space truss shown in Fig. 2.6, given that all members are made from the same uniform section, with the following material properties:

$$\rho = 2700 \text{ kg/m}^3 \qquad E = 6.7 \times 10^{10} \text{ N/m}^2$$

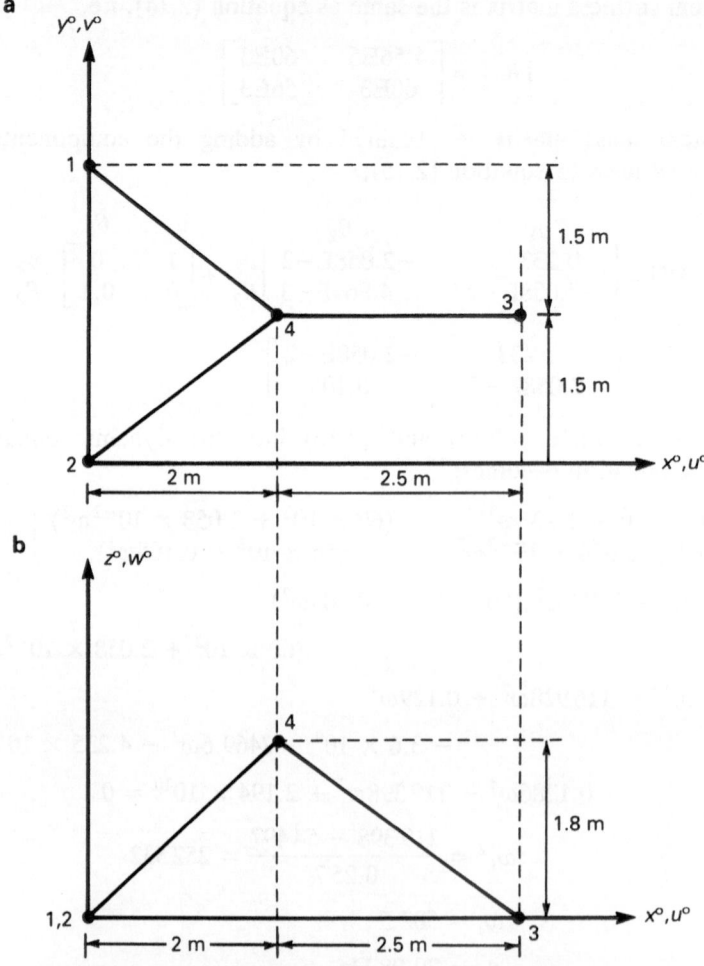

Fig. 2.6. Pin-jointed space truss. **a** Plan. **b** Front elevation.

From reference [7],

$$[k^\circ] = \left[\begin{array}{c|c} a & -a \\ \hline -a & a \end{array} \right] = \text{elemental stiffness matrix for a three-dimensional rod}$$

(2.17)

where,

$$[a] = \frac{AE}{l} \begin{array}{c} \\ \begin{bmatrix} C_{xx}{}^{\circ 2} & \text{symmetrical} & \\ C_{xx}{}^{\circ}C_{xy}{}^{\circ} & C_{xy}{}^{\circ 2} & \\ C_{xx}{}^{\circ}C_{xz}{}^{\circ} & C_{xy}{}^{\circ}C_{xz}{}^{\circ} & C_{xz}{}^{\circ 2} \end{bmatrix} \end{array} \begin{array}{c} u^\circ \\ v^\circ \\ w^\circ \end{array}$$

(2.18)

$$C_{xx}^\circ = (x_j^\circ - x_i^\circ)/l$$
$$C_{xy}^\circ = (y_j^\circ - y_i^\circ)/l \qquad\qquad (2.19)$$
$$C_{xz}^\circ = (z_j^\circ - z_i^\circ)/l$$

$x_i^\circ,\ y_i^\circ,\ z_i^\circ$ = coordinates of the "i" node

$x_j^\circ,\ y_j^\circ,\ z_j^\circ$ = coordinates of the "j" node

A = cross-sectional area

$$l = \sqrt{[(x_j^\circ - x_i^\circ)^2 + (y_j^\circ - y_i^\circ) + (z_j^\circ - z_i^\circ)^2]}$$

= element length $\qquad\qquad (2.20)$

E = Young's modulus of elasticity

$u^\circ,\ v^\circ$ and w° = displacements in the $x^\circ,\ y^\circ$ and z° directions, respectively

Also from reference [7],

$$[m^\circ] = \frac{\rho A l}{6}
\begin{bmatrix}
2 & & & & & \\
0 & 2 & & & \text{symmetrical} & \\
0 & 0 & 2 & & & \\
1 & 0 & 0 & 2 & & \\
0 & 1 & 0 & 0 & 2 & \\
0 & 0 & 1 & 0 & 0 & 2
\end{bmatrix}
\begin{matrix}
u_i^\circ \\ v_i^\circ \\ w_i^\circ \\ u_j^\circ \\ v_j^\circ \\ w_j^\circ
\end{matrix}
\qquad (2.21)$$

with column headers $u_i^\circ\quad v_i^\circ\quad w_i^\circ\quad u_j^\circ\quad v_j^\circ\quad w_j^\circ$

= elemental mass matrix for a three-dimensional rod

where,

$$\rho = \text{density}$$

Element 1–4

From equation (2.20),

$$l_{1-4} = \sqrt{[(2 - 0)^2 + (1.5 - 3)^2 + (1.8 - 0)^2]}$$
$$= 3.08 \text{ m}$$

And from equation (2.19),

$$C_{xx}^\circ = 2/3.08 = 0.649$$
$$C_{xy}^\circ = (1.5 - 3)/3.08 = -0.487$$
$$C_{xz}^\circ = (1.8 - 0)/3.08 = 0.584$$

$$[k_{1-4}^\circ] = \frac{6.7 \times 10^{10} \times A}{3.08} \begin{bmatrix} 0.421 & -0.316 & 0.379 \\ -0.316 & 0.237 & -0.284 \\ 0.379 & -0.284 & 0.341 \end{bmatrix}$$

$$[m_{1-4}^\circ] = \frac{2700 \times A \times 3.08}{6} \begin{bmatrix} 2 & 0 & 0 \\ 0 & 2 & 0 \\ 0 & 0 & 2 \end{bmatrix}$$

Element 2–4

From equations (2.20) and (2.19),

$$l_{2-4} = \sqrt{[(2 - 0)^2 + (1.5 - 0)^2 + (1.8 - 0)^2]}$$
$$= 3.08 \text{ m}$$

and,

$$C_{xx}^\circ = 0.649$$
$$C_{xy}^\circ = 0.487$$
$$C_{xz}^\circ = 0.584$$

$$[k_{2-4}^\circ] = \frac{6.7 \times 10^{10} \times A}{3.08} \begin{bmatrix} 0.421 & 0.316 & 0.379 \\ 0.316 & 0.237 & 0.284 \\ 0.379 & 0.284 & 0.341 \end{bmatrix}$$

$$[m_{2-4}^\circ] = \frac{2700 \times A \times 3.08}{6} \begin{bmatrix} 2 & 0 & 0 \\ 0 & 2 & 0 \\ 0 & 0 & 2 \end{bmatrix}$$

Element 3–4

From equations (2.20) and (2.19),

$$l_{3-4} = \sqrt{[(2 - 4.5)^2 + (1.5 - 1.5)^2 + (1.8 - 0)^2]}$$
$$= 3.08 \text{ m}$$

and,

$$C_{xx}^\circ = -0.812$$
$$C_{xy}^\circ = 0$$
$$C_{xz}^\circ = 0.584$$

$$[k_{3-4}^\circ] = \frac{6.7 \times 10^{10} \times A}{3.08} \begin{bmatrix} 0.659 & 0 & -0.474 \\ 0 & 0 & 0 \\ -0.474 & 0 & 0.341 \end{bmatrix}$$

$$[m_{3-4°}] = \frac{2700 \times A \times 3.08}{6} \begin{bmatrix} 2 & 0 & 0 \\ 0 & 2 & 0 \\ 0 & 0 & 2 \end{bmatrix}$$

The system stiffness matrix $[K°]$ and the system mass matrix $[M°]$ are given by the following:

$$[K°] = 2.175 \times 10^{10} A \begin{array}{ccc} u_4° & v_4° & w_4° \\ \begin{bmatrix} 1.501 & 0 & 0.284 \\ 0 & 0.474 & 0 \\ 0.284 & 0 & 1.023 \end{bmatrix} & \begin{array}{l} u_4° \\ v_4° \\ w_4° \end{array} \end{array}$$

$$[M°] = 1386A \begin{array}{ccc} u_4° & v_4° & w_4° \\ \begin{bmatrix} 6 & 0 & 0 \\ 0 & 6 & 0 \\ 0 & 0 & 6 \end{bmatrix} & \begin{array}{l} u_4° \\ v_4° \\ w_4° \end{array} \end{array}$$

Substituting $[K°]$ and $[M°]$ into the dynamical equation, the eigenvalue equation becomes:

$$\left| \frac{2.175 \times 10^{10} A}{6 \times 1386A} \begin{bmatrix} 1.501 & 0 & 0.284 \\ 0 & 0.474 & 0 \\ 0.284 & 0 & 1.023 \end{bmatrix} - \omega^2 \begin{bmatrix} 1 & 0 & 0 \\ 0 & 1 & 0 \\ 0 & 0 & 1 \end{bmatrix} \right| = 0$$

or

$$\begin{vmatrix} (3.926 \times 10^6 - \omega^2) & 0 & 0.743 \times 10^6 \\ 0 & (1.24 \times 10^6 - \omega^2) & 0 \\ 0.743 \times 10^6 & 0 & (2.67 \times 10^6 - \omega^2) \end{vmatrix} = 0$$

To simplify computation, interchange the first row with the second row, and the first column with the second column, as follows:

$$\begin{vmatrix} 0 & (1.24 \times 10^6 - \omega^2) & 0 \\ (3.926 \times 10^6 - \omega^2) & 0 & 0.743 \times 10^6 \\ 0.743 \times 10^6 & 0 & (2.67 \times 10^6 - \omega^2) \end{vmatrix} = 0$$

First change

$$\begin{vmatrix} (1.24 \times 10^6 - \omega^2) & 0 & 0 \\ 0 & (3.926 \times 10^6 - \omega^2) & 0.743 \times 10^6 \\ 0 & 0.743 \times 10^6 & (2.67 \times 10^6 - \omega^2) \end{vmatrix} = 0$$

Second change

Hence, the expansion of the determinant gives:

$$(1.24 \times 10^6 - \omega^2)[(3.926 \times 10^6 - \omega^2)(2.676 \times 10^6 - \omega^2)$$

$$- (0.743 \times 10^6)^2] = 0$$

By inspection,

$$\omega_1 = \sqrt{(1.24 \times 10^6)} = 1113.55$$

$$n_1 = 177.2 \text{ Hz}$$

Expanding the expression within the square brackets,

$$(3.926 \times 10^6 - \omega^2)(2.676 \times 10^6 - \omega^2) - (0.743 \times 10^6)^2 = 0$$

or,

$$1.051 \times 10^{13} - 6.602 \times 10^6 \omega^2 + \omega^4 - 5.52 \times 10^{11}$$

i.e.

$$\omega^4 = 6.602 \times 10^6 \omega^2 + 9.958 \times 10^{12}$$

Hence,

$$\omega^2 = \frac{6.602 \times 10^6 - 1.938 \times 10^6}{2}$$

or,

$$\omega_2 = 1527.1$$

$$n_2 = 243.0 \text{ Hz}$$

and,

$$\omega^2 = \frac{6.602 \times 10^6 + 1.938 \times 10^6}{2}$$

or,

$$\omega_3 = 2066$$

$$n_3 = 328.9 \text{ Hz}$$

Example 2.5 was a four-dimensional problem, and although the structure was of a very simple type, the eigenvalue solution involved the expansion of a third-order determinant. In fact, if a similar truss had two free nodes, the determinant would have been of order six, and very difficult to solve by "hand". Furthermore, Example 2.5 happened to be an uncoupled problem, and because of this, one of the eigenvalues could be directly calculated, resulting in a solution that was simpler than would normally occur with such a problem.

It is evident, therefore, that for practical structures, the determination of resonant frequencies becomes extremely difficult without the aid of a computer and, for this reason, this book has been prepared.

Chapter 3

The Modular Approach in Finite Element Programming

A flow diagram for these programs, followed by algorithms on various sections of a typical program, is presented in this chapter.

3.1 Flow Diagram

A typical flow diagram for a finite element structural vibrations' program is shown in Fig. 3.1.

3.2 The Modular Approach in Programming

From Fig. 3.1, it can be seen that after the number of elements (NELEMS) and the number of nodes (NN) are fed in, it is necessary to feed in the global coordinates of the nodes.

The global coordinates of the nodes $X(I)$ and $Y(I)$ are fed in as follows:

```
FOR I = 1 To NN
Input X(I), Y(I)
NEXT I
```

This process will store all the coordinates in two arrays, which of course should be dimensioned earlier in the program.

N.B. The above algorithm is written in BASIC, and all other algorithms will be written similarly, as this is one of the simplest of the high-level computer languages and the easiest with which to demonstrate the algorithm.

The next process is to feed in all the element details and generate and assemble the element stiffness and mass matrices. This is demonstrated by the following sub-program.

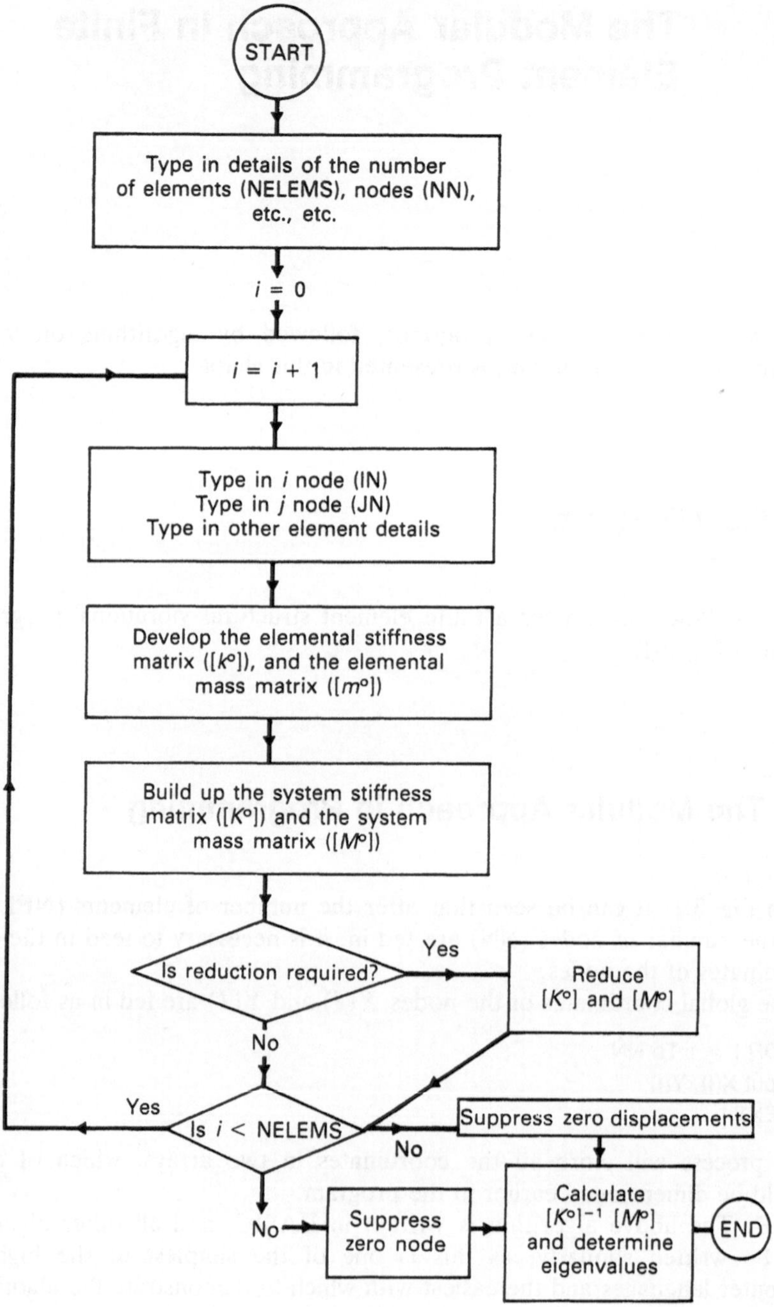

Fig. 3.1.   Flow diagram for vibration programs.

```
FOR NELEM = 1 TO NELEMS
Input I: REM the "I" node of the element
Input J: REM the "J" node of the element
Input other material and geometrical properties
L = SQR ((X(J) − X(I)) ↑ 2 + (Y(J) − Y(I)) ↑ 2):
    Rem length of element
c = (X(J) − X(I))/L: Rem cos α
s = (Y(J) − Y(I))/L: Rem sin α
```

Hence calculate the elemental $[k°]$ and $[m°]$.

To calculate the system $[K°]$ and $[M°]$, the following process is used:

```
Rem to assemble [k°] and [m°] into [K°] and [M°]
I1 = NDF*I−NDF: REM NDF = number of degrees of freedom
                                per node
J1 = NDF*J−NDF
FOR II = 1 To NDF
FOR JJ = 1 To NDF
MM = I1 + II
MN = I1 + JJ
NM = J1 + II
NN = J1 + JJ
K(MM,MN) = k(II,JJ)
M(MM,MN) = m(II,JJ)
K(MM,MN) = k(II,JJ + NDF)
M(MM,MN) = m(II,JJ + NDF)
K(NM,MN) = k(II + NDF,JJ)
M(NM,MN) = m(II + NDF,JJ)
K(NM,MN) = k(II + NDF,JJ + NDF)
M(NM,MN) = m(II + NDF,JJ + NDF)
NEXT JJ
NEXT II
NEXT ELEM
```

After assembling the stiffness and mass matrices, it is necessary to feed in the positions of the zero displacements. This is done as follows:

```
Let NFIX = number of nodes which have zero displacements.
    NDF = number of degrees of freedom per node
    FOR I = 1 To NFIX
    Input NPOS(I): Rem the node with the zero displacement or displacements
    FOR J = 1 To NDF
    Input  NSUP(NPOS(I)*NDF−NDF+J):  The details of the displacements,
                                     whether or not the displacement is
                                     zero. If the displacement is zero, type
                                     1; else type 0.

    NEXT J
    NEXT I
```

It is now necessary to suppress the zero diplacements; this is done as follows:

```
MM = 0
FOR I = 1 To NFIX
FOR J = NDF
N1 = NSUP(NPOS(I)*NDF−NDF+J)
IF N1 = 0 THEN GOTO 100
MM = MM+1
N1 = N1−MM+1
N2 = N3−MM
IF N2 < N1 THEN GOTO 200
FOR II = N1 To N2
FOR JJ = N1 To ND
K(II,JJ) = K(II+1,JJ+1)
M(II,JJ) = M(II+1,JJ+1)
NEXT JJ
NEXT II
100 IF N1 = 1 THEN GOTO 200
FOR II = 1 To N1−1
FOR JJ = N1 To N2
K(II,JJ) = K(II,JJ+1)
K(JJ,II) = K(JJ+1,II)
M(II,JJ) = M(II,JJ+1)
M(JJ,II) = MM(JJ+1,II)
NEXT JJ,II
200 END
```

The next three processes are to invert the suppressed system stiffness matrix $[K°]$ and then to premultiply $[K°]^{-1} [M°]$, prior to obtaining the eigenvalues and eigenmodes. Subroutines for these processes appear in the Appendices and will not be discussed in this chapter.

Chapter 4

Vibrations of Plane Pin-jointed Trusses

This chapter describes how to use a FORTRAN computer program for determining the resonant frequencies and eigenmodes of plane pin-jointed trusses.

4.1 Description of Program

This program, which is given in Appendix 1, is suitable for determining the natural frequencies and eigenmodes for plane pin-jointed trusses of the types shown in Fig. 4.1. These trusses are assumed to consist of one-dimensional members which possess only axial stiffness and are connected together at their joints by smooth frictionless hinges.

The elements, which are called rods, are described by two nodal points at their ends, and any possible effects due to bending or torsion are not included. At each nodal point there are two degrees of freedom ($u°$ and $v°$), making a total of four degrees of freedom per element, as shown in Fig. 4.2.

The program can analyse plane pin-jointed trusses, made from members with various cross-sectional and elastic properties. The structures can be statically determinate or statically indeterminate and the boundary conditions can be either pin-jointed or free or constrained in one direction only.

The elemental stiffness matrix in global coordinates $[k°]$ and the elemental mass matrix in global coordinates $[m°]$ are given by equations (4.1) and (4.2). It should be noted that these matrices are of order 4×4 and this size corresponds to the four degrees of freedom per element, namely, $\{u_1°,\ v_1°,\ u_2°$ and $v_2°\}$.

$$[k°] = \frac{AE}{l}
\begin{array}{c}
\begin{array}{cccc} u_1° & v_1° & u_2° & v_2° \end{array} \\
\left[
\begin{array}{cc|cc}
c^2 & cs & -c^2 & -cs \\
cs & s^2 & -cs & -s^2 \\
\hline
-c^2 & -cs & c^2 & cs \\
-cs & -s^2 & cs & s^2
\end{array}
\right]
\begin{array}{c} u_1° \\ v_1° \\ u_2° \\ v_2° \end{array}
\end{array}
\qquad (4.1)$$

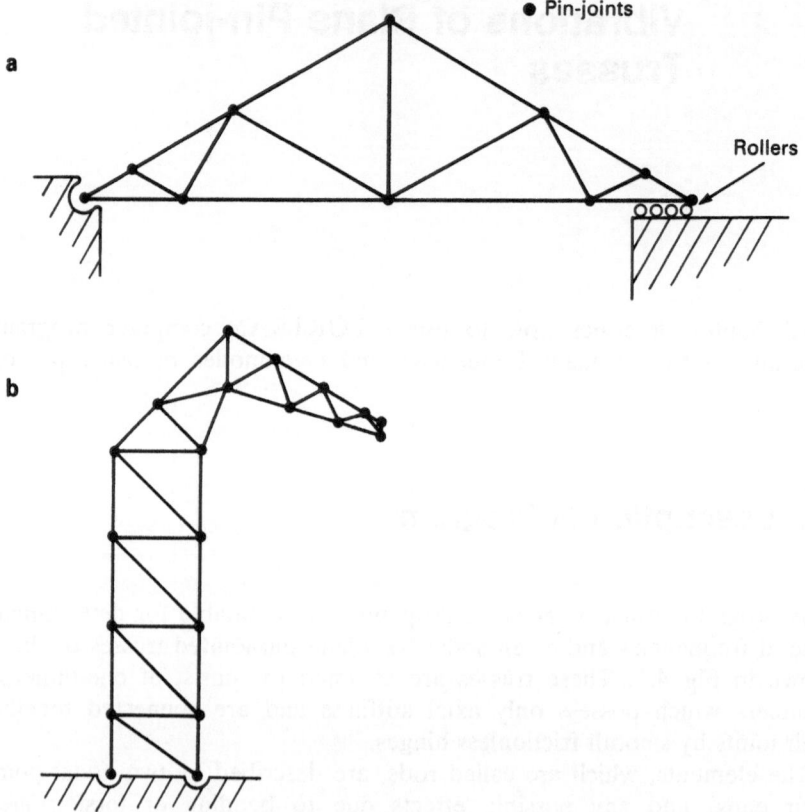

Fig. 4.1.   Pin-jointed trusses. **a** Roof truss. **b** Crane.

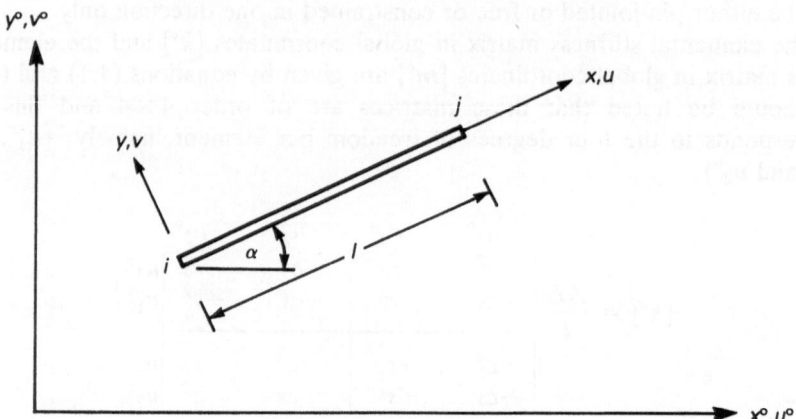

Fig. 4.2.   Rod element.

$$[m°] = \frac{\rho Al}{6} \begin{array}{cccc} \overset{u_1°}{} & \overset{v_1°}{} & \overset{u_2°}{} & \overset{v_2°}{} \\ \left[\begin{array}{cc|cc} 2 & 0 & 1 & 0 \\ 0 & 2 & 0 & 1 \\ \hline 1 & 0 & 2 & 0 \\ 0 & 1 & 0 & 2 \end{array}\right] & \begin{array}{c} u_1° \\ v_1° \\ u_2° \\ v_2° \end{array} \end{array}$$

(4.2)

where,

   A = cross-sectional area

   l = elemental length

   ρ = density of element material

   E = Young's modulus of elasticity of element material

   $c = \cos(\alpha)$

   $s = \sin(\alpha)$

   α = angle of inclination of element with the $x°$ axis (see Fig. 4.2)

4.2 Data

The data should be typed into the file MAIN6DAT.DAT, as follows:

NN	LEMS	NFIXD	NV
↑	↑	↑	↑
number of nodes	number of members	number of nodes with zero displacements	number of frequencies

Details of Zero Displacements

   $i = 1(1)$NFIXD

   $NPOSS_i$ = the node with the zero displacements

   If the displacement in the $x°$ direction at the node $NPOSS_i$ is zero, type 1; else type 0 (i.e. $NXSUP_i = 1$ or 0).

   If the displacement in the $y°$ direction at the node $NPOSS_i$ is zero, type 1; else type 0 (i.e. $NYSUP_i = 1$ or 0).

   REPEAT

Global Coordinates of Nodes

> $i = 1(1)\text{NN}$
>
> Type in x_i° and y_i°,
>
> where "i" corresponds to the appropriate node
>
> ↳ REPEAT

Element Topology

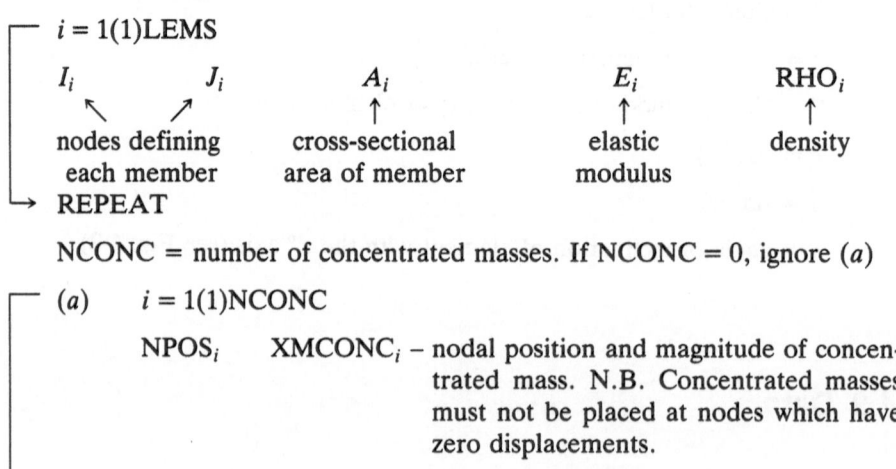

> $i = 1(1)\text{LEMS}$
>
> | I_i | J_i | A_i | E_i | RHO$_i$ |
>
> nodes defining   cross-sectional   elastic   density
> each member   area of member   modulus
> ↳ REPEAT

NCONC = number of concentrated masses. If NCONC = 0, ignore (*a*)

> (*a*) $i = 1(1)\text{NCONC}$
>
>   NPOS$_i$ XMCONC$_i$ – nodal position and magnitude of concentrated mass. N.B. Concentrated masses must not be placed at nodes which have zero displacements.
>
>   REPEAT

Output

The output, which will appear in the file OUT6.OUT, will be as follows:

Details of input data, plus,

> $i = 1(1)\text{NV}$
>
> Frequency (n_i)
>
> ↳ Eigenmode $[u^\circ \quad v^\circ]_i$

4.3 Vibration Problems Involving Plane Pin-jointed Trusses

Example 4.1

Determine the two lowest natural frequencies of vibration for the plane pin-jointed truss shown in Fig. 4.3.

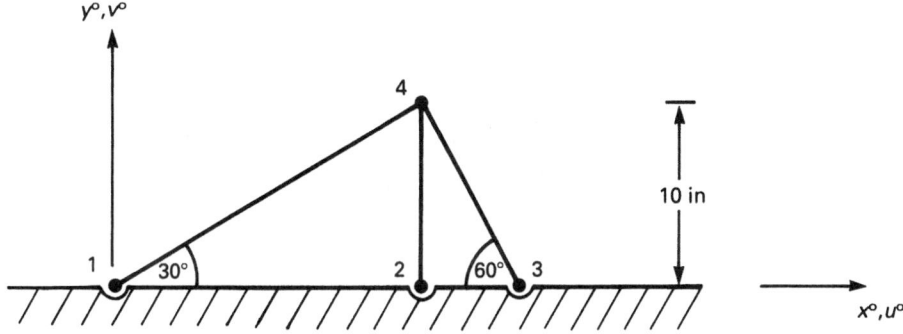

Fig. 4.3.

All members of the truss may be assumed to have the following sectional and material properties:

cross-sectional area $= 0.06$ in^2

elastic modulus $\quad = 30 \times 10^6$ lbf/in^2

density $\quad\quad\quad = 7.35 \times 10^{-4}$ lbf.s^2/in^4

See Appendices A1.2 and A1.3. The data for this example are as follows:

4	3	3	2
↑	↑	↑	↑
NN	LEMS	NFIXD	NV

Details of Zero Displacements

NPOSS	NXSUP	NYSUP
1	1	1
2	1	1
3	1	1

Global Coordinates of Nodes

$x_i{}^\circ$	$y_i{}^\circ$
0.00	0.0
17.32	0.0
23.09	0.0
17.32	10.0

Element Particulars

i	j	A	E	ρ
1	4	0.06	30×10^6	7.35×10^{-4}
2	4	0.06	30×10^6	7.35×10^{-4}
3	4	0.06	30×10^6	7.35×10^{-4}

$\text{NCONC} = 0$

Results

The results for this particular example are:

First eigenvalue $\quad = \lambda_1 = 5.948 \times 10^{-9}$

First frequency $\quad = n_1 = \sqrt{(1/\lambda_1)}/2\pi = 2064 \text{ Hz}$

Node	u°	v°
1	0.0	0.0
2	0.0	0.0
3	0.0	0.0
4	1.0	0.132

This is the first eigenmode – see Fig. 4.4.

Second frequency $= n_2 = 3662 \text{ Hz}$

Node	u°	v°
1	0.0	0.0
2	0.0	0.0
3	0.0	0.0
4	−0.132	1.0

This is the second eigenmode – see Fig. 4.4.

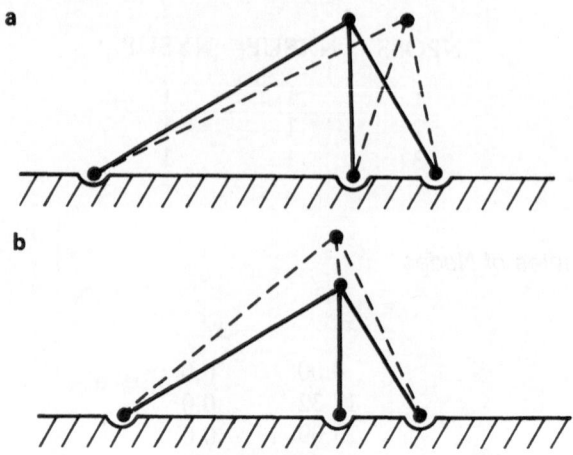

Fig. 4.4.   a First eigenmode. **b** Second eigenmode.

Example 4.2

Determine the three lowest natural frequencies of vibration for the plane pin-jointed truss shown in Fig. 4.5.

$$\text{For all members except 3–4,} \quad A = 9.7 \times 10^{-4} \text{ m}^2$$

$$\text{for member 3–4,} \quad A = 1.2 \times 10^{-3} \text{ m}^2$$

$$\text{For all members,} \quad E = 2 \times 10^{11} \text{ N/m}^2$$

$$\rho = 7860 \text{ kg/m}^3$$

Data

The input file for this problem will be as follows:

6	9	2	3
1	1	1	
6	0	1	
0	0		
3	1.5		
6	0		
6	3		
9	1.5		
12	0		
1	2	9.7E–4	2E11	7860
1	3	9.7E–4	2E11	7860
2	4	9.7E–4	2E11	7860
2	3	9.7E–4	2E11	7860
4	3	9.7E–4	2E11	7860
4	5	9.7E–4	2E11	7860
3	5	9.7E–4	2E11	7860
3	6	9.7E–4	2E11	7860
6	5	9.7E–4	2E11	7860
0				

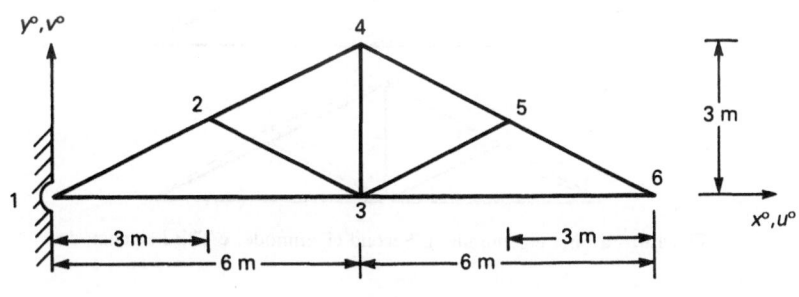

Fig. 4.5.

Results

$$\left.\begin{array}{l} \lambda_1 = 2.74 \times 10^{-5} \\ \lambda_2 = 3.89 \times 10^{-6} \\ \lambda_3 = 2.04 \times 10^{-6} \end{array}\right\} \text{eigenvalues}$$

1st frequency $= n_1 = \sqrt{(1/\lambda_1)}/2\pi = 30.4$ Hz

Node	u°	v°
1	0.0	0.0
2	−0.328	0.907
3	−0.252	1.0
4	−0.246	0.954
5	−0.171	0.878
6	−0.469	0.0

$\left.\right\}$ 1st eigenmode – see Fig. 4.6

2nd frequency $= n_2 = 80.7$ Hz

Node	u°	v°
1	0.0	0.0
2	0.596	−0.453
3	0.938	0.443
4	0.565	0.318
5	1.0	0.770
6	0.851	0.0

$\left.\right\}$ 2nd eigenmode – see Fig. 4.6

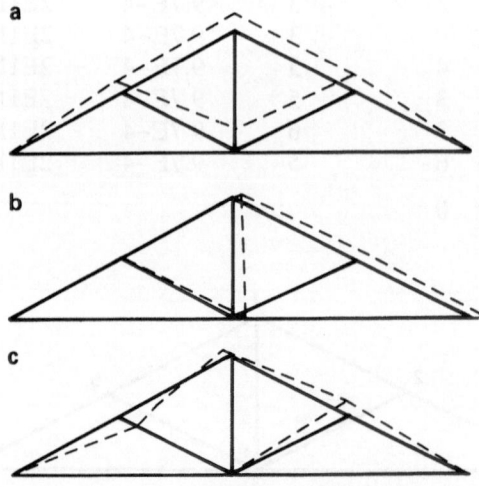

Fig. 4.6. a First eigenmode. **b** Second eigenmode. **c** Third eigenmode.

3rd frequency $= n_3 = 111.6\,\text{Hz}$

Node	$u°$	$v°$	
1	0.0	0.0	
2	−0.019	−0.987	
3	0.153	−0.12	3rd eigenmode – see Fig. 4.6
4	−0.7	−0.063	
5	−0.19	1.0	
6	−0.401	0.0	

The results from Examples 4.1 and 4.2 compare favourably with those from [8], although some loss of precision has been made in the current text for increase in speed of calculation.

Example 4.3

If the truss of Example 4.2 has concentrated masses of magnitude 2 kg, 4 kg and 3 kg placed at nodes 3, 4 and 5, respectively, determine the natural frequencies of vibration.

Data

The data are as for Example 4.2, except for the following:

$$\text{NCONC} = 3, \text{ plus,}$$

3	2	
4	4	details of nodes with the additional concentrated masses
5	3	

Results

Eigenvalues

$$\lambda_1 = 2.868 \times 10^{-5}$$

$$\lambda_2 = 4.064 \times 10^{-6}$$

$$\lambda_3 = 2.185 \times 10^{-6}$$

1st frequency = n_1 = 29.72 Hz

Node	u°	v°
1	0.000	0.000
2	−0.326	0.903
3	−0.250	1.000
4	−0.247	0.955
5	−0.169	0.883
6	−0.468	0.000

$\left. \right\}$ 1st eigenmode

2nd frequency = n_2 = 78.95 Hz

Node	u°	v°
1	0.000	0.000
2	0.597	−0.465
3	0.936	0.421
4	0.568	0.299
5	1.000	0.760
6	0.849	0.000

$\left. \right\}$ 2nd eigenmode

3rd frequency = n_3 = 107.67 Hz

Node	u°	v°
1	0.000	0.000
2	−0.018	−0.991
3	0.158	−0.145
4	−0.709	−0.088
5	−0.188	1.000
6	−0.392	0.000

$\left. \right\}$ 3rd eigenmode

Chapter 5

Vibrations of Continuous Beams

This chapter describes how to use a FORTRAN computer program for determining the resonant frequencies and eigenmodes for encastré and continuous beams.

5.1 Description of Program

This program, which is given in Appendix 2, is suitable for determining the natural frequencies of vibration of straight beams with one or more supports. The beams can be in the form of cantilevers or they can be simply-supported or hinged at two or more supports, as shown in Fig. 5.1. The boundary conditions can be of various combinations, including fixed and hinged, or they can be of the type that restricts motion in one direction only.

The beam can be of the type that is manufactured from several elements of different sectional properties, and it is a simple matter to allow the elements to have different material properties.

The beam is assumed to consist of a number of horizontal elements connected to each other at the nodal points, as shown in Fig. 5.1. From Fig. 5.1, it can be seen that there are eight elements and nine nodes. Node 1 is fixed and node 9 is hinged and vertical movement is restricted at nodes 2, 3, 5, 7 and 8. Nodes 4 and 6 have been chosen because they lie at discontinuities, and if desired, additional points could have also been chosen as nodes.

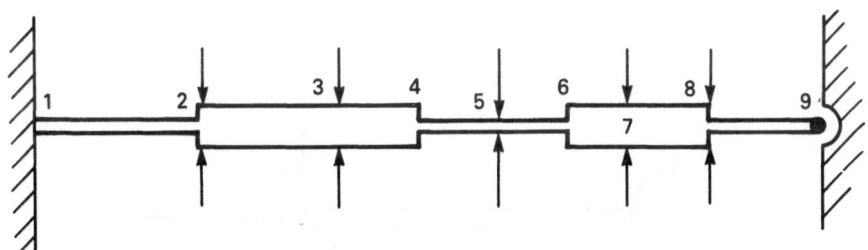

Fig. 5.1. Continuous beam.

Each nodal point has two degrees of freedom per node (v and θ), as shown in Fig. 5.2.

The elemental stiffness matrix $[k]$ and the elemental mass matrix $[m]$ are given in equations (5.1) and (5.2). These matrices are of order 4×4 and the reason for this is that they correspond to the four degrees of freedom per element, namely, $\{v_1, \theta_1, v_2, \theta_2\}$.

$[k]$ = the elemental stiffness matrix for a beam

$$= EI \begin{array}{cccc} v_1 & \theta_1 & v_2 & \theta_2 \\ \begin{bmatrix} 12/l^3 & -6/l^2 & -12/l^3 & -6/l^2 \\ -6/l^2 & 4/l & 6/l^2 & 2/l \\ -12/l^3 & 6/l^2 & 12/l^3 & 6/l^2 \\ -6/l^2 & 2/l & 6/l^2 & 4/l \end{bmatrix} & \begin{array}{c} v_1 \\ \theta_1 \\ v_2 \\ \theta_2 \end{array} \end{array} \tag{5.1}$$

$[m]$ = the elemental mass matrix for a beam

$$= \frac{\rho Al}{420} \begin{array}{cccc} v_1 & \theta_1 & v_2 & \theta_2 \\ \begin{bmatrix} 156 & \text{symmetrical} & & \\ -22l & 4l^2 & & \\ 54 & -13l & 156 & \\ 13l & -3l^2 & 22l & 4l^2 \end{bmatrix} & \begin{array}{c} v_1 \\ \theta_1 \\ v_2 \\ \theta_2 \end{array} \end{array} \tag{5.2}$$

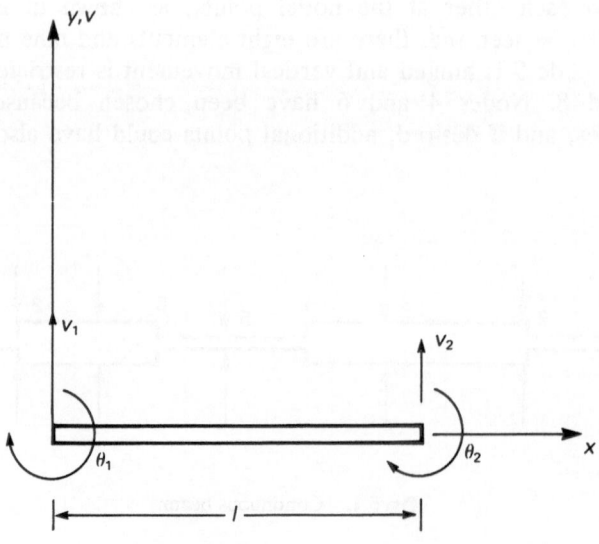

Fig. 5.2. Beam element.

5.2 Data

The data should be typed into the file MAIN7DAT.DAT, as follows:

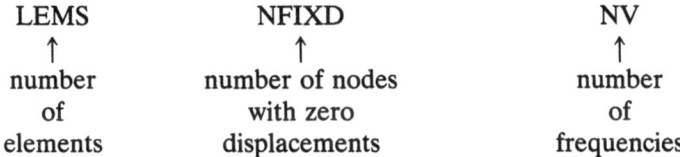

LEMS   NFIXD   NV
↑   ↑   ↑

| number of elements | number of nodes with zero displacements | number of frequencies |

Details of Zero Displacements

$i = 1(1)$NFIXD

$NPOS_i$ = the node with the zero diplacements

If $v(NPOS_i) = 0$, type 1; else type 0 (zero)

If $\theta(NPOS_i) = 0$, type 1; else type 0 (zero)

↳ REPEAT

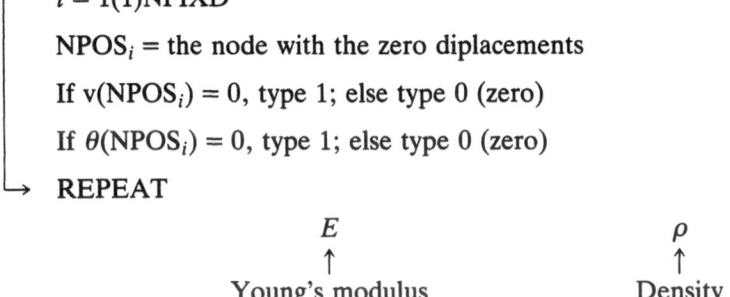

E ρ
↑   ↑
Young's modulus   Density

Element Topology

$i = 1(1)$LEMS

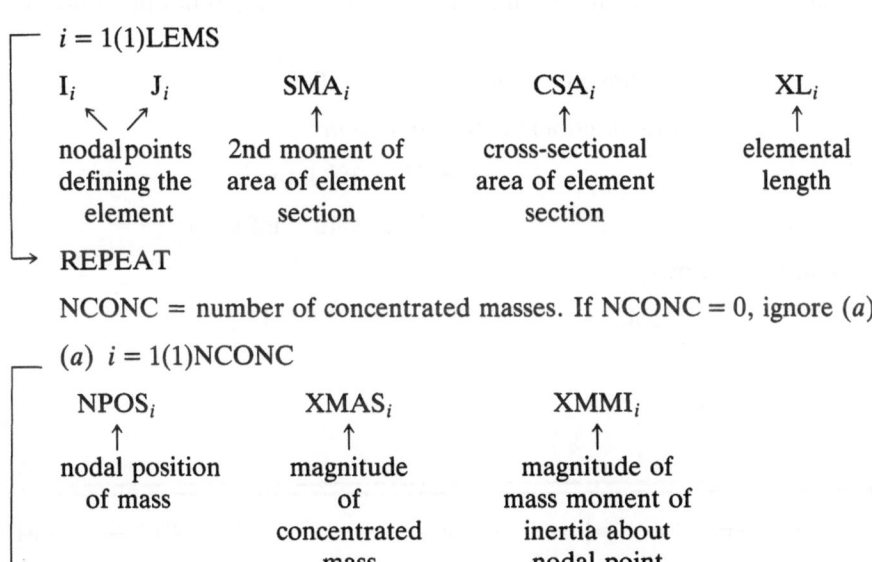

I_i J_i SMA_i CSA_i XL_i

| nodal points defining the element | 2nd moment of area of element section | cross-sectional area of element section | elemental length |

↳ REPEAT

NCONC = number of concentrated masses. If NCONC = 0, ignore (*a*)

(*a*) $i = 1(1)$NCONC

$NPOS_i$ $XMAS_i$ $XMMI_i$
↑   ↑   ↑

| nodal position of mass | magnitude of concentrated mass | magnitude of mass moment of inertia about nodal point |

↳ REPEAT

N.B. Concentrated masses should not be placed at nodes with suppressed displacements, and *element details should always be fed in from left to right of the beam*.

Results

The output should appear on the file OUT7.OUT and should include the following:

Details of input, plus,

$$
\left[
\begin{array}{l}
i = 1(1)\text{NV} \\
\text{Frequencies} = n_i \\
j = 1(1)\ \text{LEMS} + 1 \\
v_j \quad \theta_j - \text{eigenmodes} \\
\longrightarrow \text{REPEAT}
\end{array}
\right.
$$

5.3 Continuous Beams Problems

Example 5.1

Determine the two lowest natural frequencies of vibration for the cantilever beam shown in Fig. 5.3. The beam is of constant section and has the following particulars:

$$\text{Second moment of area} = 3.25 \times 10^{-4} \text{ in}^4$$

$$\text{Cross-sectional area} = 0.0625 \text{ in}^2$$

$$E = 30 \times 10^6 \text{ lbf/in}^2$$

$$\rho = 7.35 \times 10^{-4} \text{ lbf.s}^2/\text{in}^4$$

The input is as follows:

$$\text{LEMS} = 4 \qquad \text{NFIXD} = 1 \qquad \text{NV} = 2$$

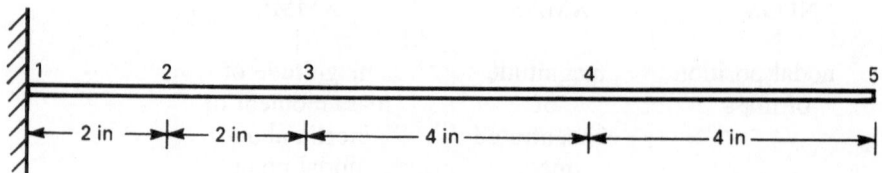

Fig. 5.3.   Cantilever beam.

Details of Zero Displacements

$$1 \quad 1 \quad 1$$

$$E = 30 \times 10^6 \qquad RHO = 7.35 \times 10^{-4}$$

Element Particulars

i	j	I	A	Length
1	2	3.25×10^{-4}	0.0625	2.0
2	3	3.25×10^{-4}	0.0625	2.0
3	4	3.25×10^{-4}	0.0625	4.0
4	5	3.25×10^{-4}	0.0625	4.0

$$NCONC = 0$$

The *results* are as follows:

Eigenvalues

$$\lambda_1 = 7.9 \times 10^{-6}$$
$$\lambda_2 = 2.0 \times 10^{-7}$$

1st frequency = n_1 = 56.6 Hz

Node v		θ	
1	0.0	0.0	
2	0.045	−0.043	
3	0.166	−0.075	1st eigenmode – see Fig. 5.4
4	0.547	−0.109	
5	1.0	−0.115	

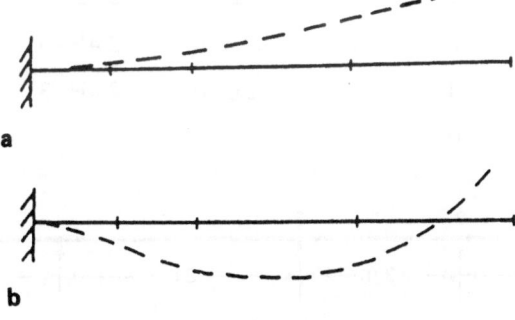

Fig. 5.4. Eigenmodes of cantilever beam. **a** First eigenmode. **b** Second eigenmode.

2nd frequency = n_2 = 355.8 Hz

Node	v	θ	
1	0.0	0.0	
2	−0.225	0.185	
3	−0.59	0.147	2nd eigenmode – see Fig. 5.4
4	−0.423	−0.247	
5	1.0	−0.399	

Example 5.2

Determine the three lowest natural frequencies of vibration for the uniform section continuous beam shown in Fig. 5.5.

The following particulars apply to the beam:

$$I = 4.2 \times 10^{-6} \text{ m}^4$$

$$A = 2.4 \times 10^{-3} \text{ m}^2$$

$$E = 2 \times 10^{11} \text{ N/m}^2$$

$$\rho = 7860 \text{ kg/m}^3$$

Data

The input file for this problem is as follows:

4	4	3		
1	1	1		
3	1	0		
4	1	0		
5	1	0		
1	2	4.2E–6	2.4E–3	2
2	3	4.2E–6	2.4E–3	2
3	4	4.2E–6	2.4E–3	3
4	5	4.2E–6	2.4E–3	3
0				

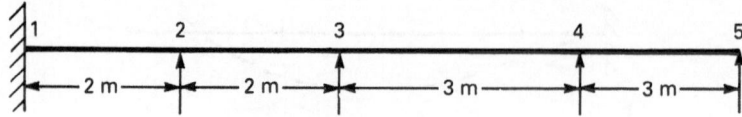

Fig. 5.5.   Continuous beam.

Results

Eigenvalues

$$\lambda_1 = 1.848 \times 10^{-5}$$
$$\lambda_2 = 5.062 \times 10^{-6}$$
$$\lambda_3 = 1.625 \times 10^{-6}$$

Frequencies and Eigenmodes

$n_1 = 37.02$ Hz

Node v		θ	
1	0.0	0.0	
2	1.0	−0.203	
3	0.0	0.724	1st eigenmode – see Fig. 5.6
4	0.0	−0.280	
5	0.0	0.157	

$n_2 = 70.74$ Hz

Node v		θ	
1	0.0	0.0	
2	0.341	0.225	
3	0.0	−0.597	2nd eigenmode
4	0.0	1.0	
5	0.0	−0.754	

$n_3 = 124.9$ Hz

Node v		θ	
1	0.0	0.0	
2	0.136	0.705	
3	0.0	−0.740	3rd eigenmode
4	0.0	−0.531	
5	0.0	1.0	

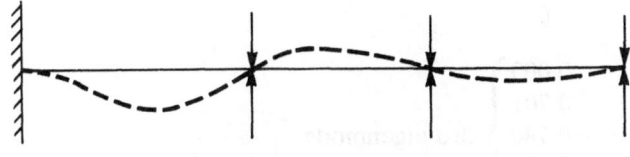

Fig. 5.6. First eigenmode.

Example 5.3

Determine the three lowest natural frequencies of vibration for the beam of Example 5.2, assuming that it carries an additional mass of 1.5 kg at node 2, together with a mass moment of inertia of 1.09×10^{-3} kg m^2 at this node. See Appendices A2.2 and A2.3.

Data

The data are as for Example 5.2, except for the following concerning the added mass:

$$\text{NCONC} = 1 \text{ instead of } 0, \text{ plus,}$$

2 1.5 0.00109 (nodal position of mass, together with the mass and mass moment of inertia)

Results

1st frequency = n_1 = 36.3 Hz

Node	v	θ	
1	0.000	0.000	
2	1.000	−0.199	
3	0.000	0.713	1st eigenmode
4	0.000	−0.270	
5	0.000	0.151	

2nd frequency = n_2 = 70.42 Hz

Node	v	θ	
1	0.000	0.000	
2	0.322	0.230	
3	0.000	−0.612	2nd eigenmode
4	0.000	1.000	
5	0.000	−0.752	

3rd frequency = n_3 = 124.6 Hz

Node	v	θ	
1	0.000	0.000	
2	0.128	0.701	
3	0.000	−0.740	3rd eigenmode
4	0.000	−0.534	
5	0.000	1.000	

Chapter 6

Vibrations of Rigid-jointed Plane Frames

This chapter describes how to use a FORTRAN computer program for determining the resonant frequencies and eigenmodes of rigid-jointed plane frames.

6.1 Description of Program

This program, which is given in Appendix 3, is intended for determining the natural frequencies of vibration of rigid-jointed plane frames of the type shown in Fig. 6.1. These frames can be portal or skew, and if required, they can be multi-bay.

The boundary conditions can be fixed or pinned or they can be restrained in one direction only.

The frames can be made from members of different sectional properties and it is a simple matter to extend the program to cater for members with different material properties.

Each member or element has two nodal points and it is at these points that the element is connected to other elements. There are three degrees of

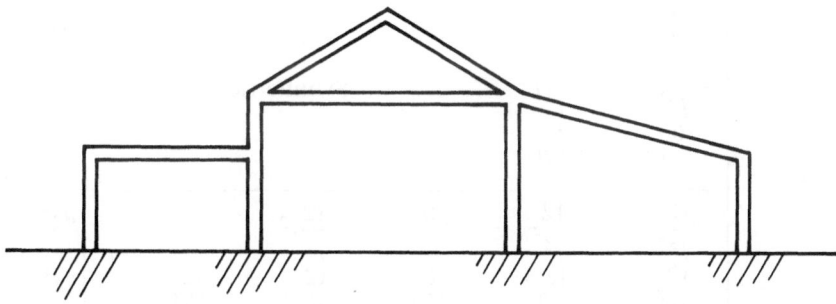

Fig. 6.1. Multi-bay frame.

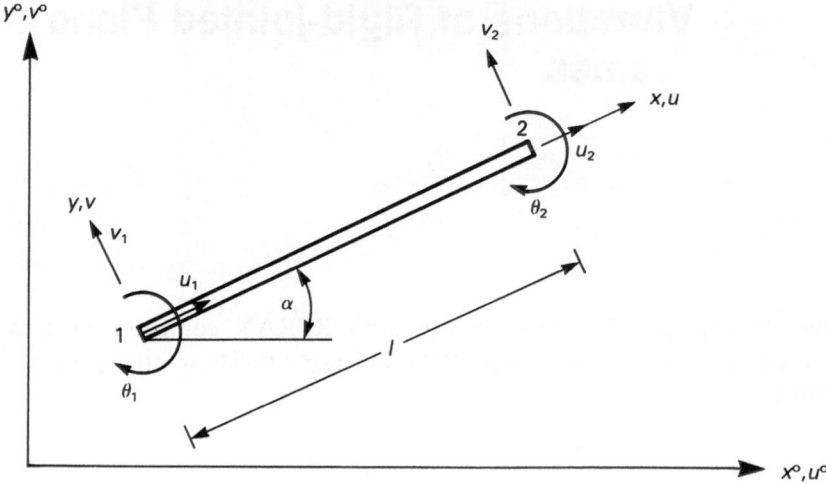

Fig. 6.2.   Beam element.

freedom at each node ($u°$, $v°$ and θ), as shown in Fig. 6.2, making a total of six degrees of freedom per element.

The element stiffness matrix $[k°]$ is obtained by superimposing the stiffness matrix of the rod element of Chap. 4 with the beam element of Chap. 5. The mass matrix, as shown by equation (6.2), is obtained in a similar manner, but the process is not quite as straightforward and is discussed more fully in [7].

$$[k°] = [k_b°] + [k_r°]$$

where

$$[k_b°] = EI
\begin{bmatrix}
\frac{12}{l^3}s^2 & & & & & \\
-\frac{12}{l^3}cs & \frac{12}{l^3}c^2 & & \text{Symmetrical} & & \\
\frac{6}{l^2}s & -\frac{6}{l^2}c & \frac{4}{l} & & & \\
-\frac{12}{l^3}s^2 & \frac{12}{l^3}cs & -\frac{6}{l^2}s & \frac{12}{l^3}s^2 & & \\
\frac{12}{l^3}cs & -\frac{12}{l^3}c^2 & \frac{6}{l^2}c & -\frac{12}{l^3}cs & \frac{12}{l^3}c^2 & \\
\frac{6}{l^2}s & -\frac{6}{l^2}c & \frac{2}{l} & \frac{6}{l^2}s & -\frac{6}{l^2}c & \frac{4}{l}
\end{bmatrix}
\begin{matrix}
u_1° \\ v_1° \\ \theta_1 \\ u_2° \\ v_2° \\ \theta_2
\end{matrix}
\qquad (6.1)$$

with column headings $u_1°$, $v_1°$, θ_1, $u_2°$, $v_2°$, θ_2

$$[k_r{}^\circ] = \frac{AE}{l} \begin{bmatrix} \overset{u_1{}^\circ}{c^2} & \overset{v_1{}^\circ}{cs} & \overset{\theta_1}{0} & \overset{u_2{}^\circ}{-c^2} & \overset{v_2{}^\circ}{-cs} & \overset{\theta_2}{0} \\ cs & s^2 & 0 & -cs & -s^2 & 0 \\ 0 & 0 & 0 & 0 & 0 & 0 \\ -c^2 & -cs & 0 & c^2 & cs & 0 \\ -cs & -s^2 & 0 & cs & s^2 & 0 \\ 0 & 0 & 0 & 0 & 0 & 0 \end{bmatrix} \begin{matrix} u_1{}^\circ \\ v_1{}^\circ \\ \theta_1 \\ u_1{}^\circ \\ v_2{}^\circ \\ \theta_2 \end{matrix}$$

= stiffness matrix for an inclined rod element.

$$[m^\circ] = [m_r]^\circ + [m_b{}^\circ] \tag{6.2}$$

where

$$[m_r{}^\circ] = \frac{\rho A l}{6} \begin{bmatrix} \overset{u_1{}^\circ}{2c^2} & \overset{v_1{}^\circ}{} & \overset{\theta_1}{} & \overset{u_2{}^\circ}{} & \overset{v_2{}^\circ}{} & \overset{\theta_2}{} \\ 2cs & 2s^2 & & \text{Symmetrical} & & \\ 0 & 0 & 0 & & & \\ c^2 & cs & 0 & 2c^2 & & \\ cs & s^2 & 0 & 2cs & 2s^2 & \\ 0 & 0 & 0 & 0 & 0 & 0 \end{bmatrix} \begin{matrix} u_1{}^\circ \\ v_1{}^\circ \\ \theta_1 \\ u_2{}^\circ \\ v_2{}^\circ \\ \theta_2 \end{matrix}$$

$$[m_b{}^\circ] = \frac{\rho A l}{420} \begin{bmatrix} \overset{u_1{}^\circ}{156S^2} & \overset{v_1{}^\circ}{} & \overset{\theta_1}{} & \overset{u_2{}^\circ}{} & \overset{v_2{}^\circ}{} & \overset{\theta_2}{} \\ -156CS & 156C^2 & & \text{Symmetrical} & & \\ 22lS & -22lC & 4l^2 & & & \\ 54S^2 & -54CS & 13lS & 156S^2 & & \\ -54CS & 54C^2 & -13lC & -156CS & 156C^2 & \\ -13lS & 13lC & -3l^2 & -22lS & 22lC & 4l^2 \end{bmatrix} \begin{matrix} u_1{}^\circ \\ v_1{}^\circ \\ \theta_1 \\ u_2{}^\circ \\ v_2{}^\circ \\ \theta_2 \end{matrix}$$

where the symbols have the same meanings as in Chaps. 4 and 5.

It should be noted that these elements are of order 6×6 and this size corresponds to the six degrees of freedom per element, namely $\{u_1{}^\circ, v_1{}^\circ, \theta_1, u_2{}^\circ, v_2{}^\circ, \theta_2\}$.

6.2 Data

The data should be typed into file MAIN8DAT.DAT, as follows:

NN	MEMS	NFIXD	NV
↑	↑	↑	↑
number of nodes	number of elements	number of nodes with zero displacements	number of frequencies

Global Coordinates of Nodes

$$\begin{array}{l} i = 1(1)\text{NN} \\ x_i^{\circ} \quad y_i^{\circ}, \\ \text{REPEAT} \end{array}$$

Details of Zero Displacements

$i = 1(1)$NFIXD

NPOSS_i = the node with the zero displacements

$\text{NXSUP}_i = 1$ if the u° displacement is zero at this node, else type in 0

$\text{NYSUP}_i = 1$ if the v° displacement is zero at this node, else type in 0

$\text{NTSUP}_i = 1$ if the θ displacement is zero at this node, else type in 0

REPEAT

E = Young's modulus ρ = density

Element Topology

$i = 1(1)$MEMS

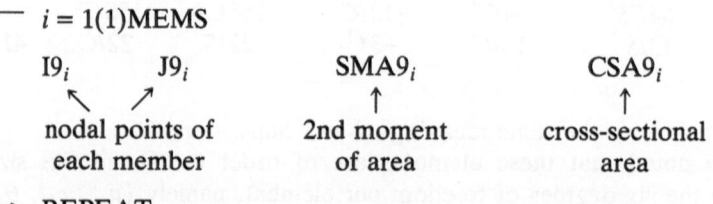

| I9_i | J9_i | SMA9_i | CSA9_i |
| nodal points of each member | | 2nd moment of area | cross-sectional area |

REPEAT

NCONC = number of concentrated masses. If NCONC = 0, ignore (a)

(a) $i = 1(1)$NCONC

NPOS_i	XMAS_i	XMMI_i
nodal position of concentrated mass	magnitude of concentrated mass	magnitude of mass moment of inertia about nodal point in the x°–y° plane

REPEAT

Results

The output file should be called OUT8.OUT, and this will include the input data from the input file, plus the following:

$i = 1(1)\text{NV}$

Frequency $= n_i$

$j = 1(1)\text{NN}$

$u_j^\circ \quad v_j^\circ \quad \theta_j$ – eigenmodes corresponding to "n_i"

6.3 Rigid-jointed Plane Frame Problems

Example 6.1

Determine the two lowest natural frequencies of vibration for the portal frame shown in Fig. 6.3. The members of the framework have the following properties:

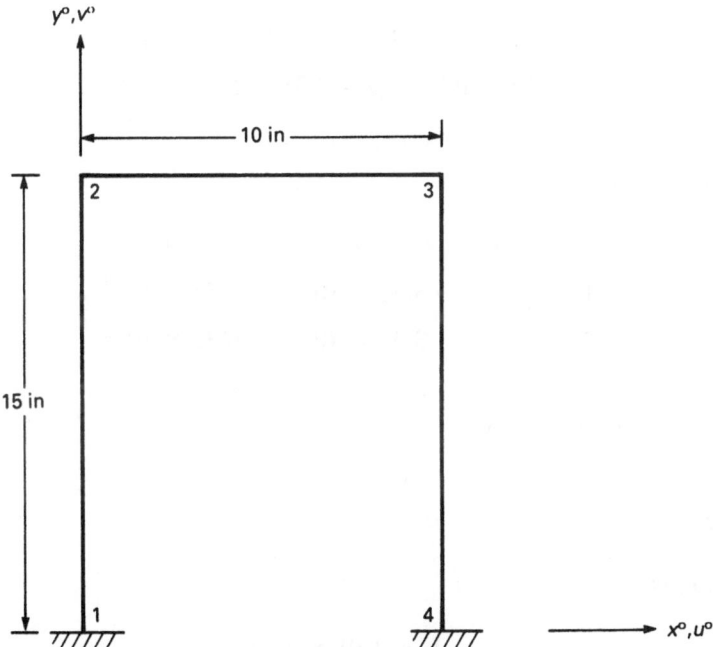

Fig. 6.3.

$$I = 3.36 \times 10^{-4} \text{ in}^4$$
$$A = 6.35 \times 10^{-2} \text{ in}^2$$
$$E = 30 \times 10^6 \text{ lbf/in}^2$$
$$\rho = 7.35 \times 10^{-4} \text{ lbf.s}^2/\text{in}^4$$

The data input is as follows:

$$\text{NN} = 4 \quad \text{MEMS} = 3 \quad \text{NFIXD} = 6 \quad \text{NV} = 2$$

Global Coordinates of Nodes

x°	y°
0.0	0.0
0.0	15.0
12.0	15.0
12.0	0.0

Details of Zero Displacements

NPOSS	NXSUP	NYSUP	NTSUP
1	1	1	1
4	1	1	1

$$E = 30 \times 10^6 \qquad \rho = 7.35 \times 10^{-4}$$

Element Details

i	j	I	A
1	2	3.36×10^{-4}	6.35×10^{-2}
2	3	3.36×10^{-4}	6.35×10^{-2}
3	4	3.36×10^{-4}	6.35×10^{-2}

$$\text{NCONC} = 0$$

Results

Eigenvalues

$$\lambda_1 = 1.91 \times 10^{-5}$$
$$\lambda_2 = 6.52 \times 10^{-7}$$

Frequencies and Eigenmodes

First frequency = n_1 = 36.4 Hz

Node	$u°$	$v°$	θ	
1	0.0	0.0	0.0	
2	1.0	2.06×10^{-4}	0.031	1st eigenmode – see Fig. 6.4a
3	1.0	-2.06×10^{-4}	0.031	
4	0.0	0.0	0.0	

Second frequency = n_2 = 197.1 Hz

Node	$u°$	$v°$	θ	
1	0.0	0.0	0.0	
2	0.019	0.0068	-1.0	2nd eigenmode – see Fig. 6.4b
3	-0.0018	0.0068	1.0	
4	0.0	0.0	0.0	

Example 6.2

Determine the two lowest natural frequencies of vibration for the skew frame shown in Fig. 6.5. The following particulars apply to the three elements of the frame:

$$I_{1-2} = 2.7 \times 10^{-6} \text{ m}^4 \qquad I_{2-3} = I_{3-4} = 1.8 \times 10^{-6} \text{ m}^4$$

$$A_{1-2} = 0.0014 \text{ m}^2 \qquad A_{2-3} = A_{3-4} = 0.0011 \text{ m}^2$$

$$E = 2 \times 10^{11} \text{ N/m}^2 \qquad \rho = 7860 \text{ kg/m}^3$$

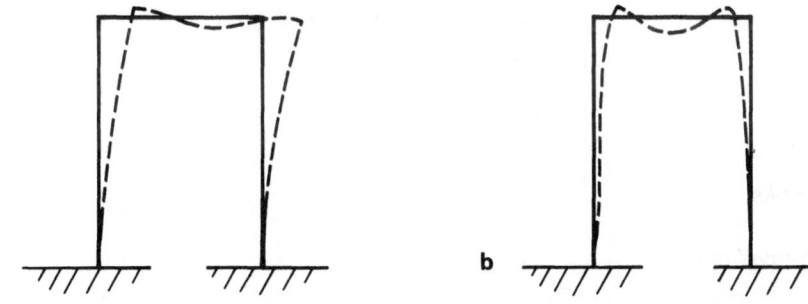

Fig. 6.4. Eigenmodes for portal frame. a First eigenmode. b Second eigenmode.

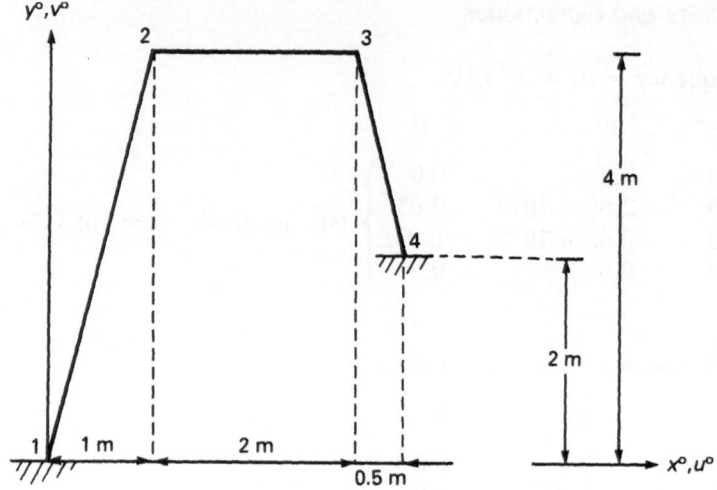

Fig. 6.5. Skew frame.

Data

The data should be typed into a file named MAIN8DAT.DAT, as follows:

4	3	2	2
0	0		
1	4		
3	4		
3.5	2		
1	1	1	1
4	1	1	1
1	2	2.7E−6	0.0014
2	3	1.8E−6	0.0011
3	4	1.8E−6	0.0011
2E11	7860		
0			

Results

Eigenvalues

$$\lambda_1 = 6.67 \times 10^{-5}$$
$$\lambda_2 = 5.10 \times 10^{-6}$$

Frequencies and Eigenmodes

First frequency = n_1 = 19.49 Hz

Node	u°	v°	θ
1	0.000	0.000	0.000
2	1.000	−0.247	−0.271
3	0.998	0.247	0.235
4	0.000	0.000	0.000

1st eigenmode – see Fig. 6.6

Second frequency = n_2 = 70.45 Hz

Node	u°	v°	θ
1	0.000	0.000	0.000
2	0.258	−0.063	1.000
3	0.258	0.059	−0.418
4	0.000	0.000	0.000

2nd eigenmode

Example 6.3

Determine the two lowest natural frequencies of vibration for the rigid-jointed plane frame of Example 6.2, but which carries the following two additional masses:

At node 2, mass = 5 kg and mass moment of inertia = 0.0144 kg.m^2

At node 3, mass = 8 kg and mass moment of inertia = 0.0523 kg.m^2

See Appendices A3.2 and A3.3. The data are the same as that for Example 6.2, except that,

NCONC = 2, plus the following:

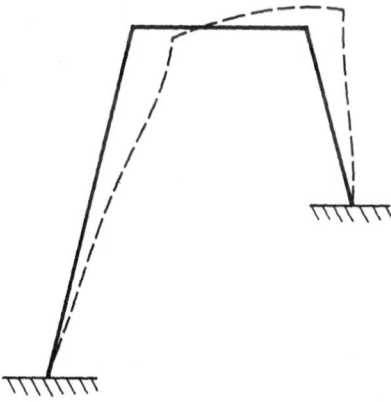

Fig. 6.6.   First eigenmode.

2	5	0.0144 ⎤	Details of nodes with additional concentrated masses and
3	8	0.0523 ⎦	mass moments of inertia

Results

$$n_1 = 17.16 \text{ Hz}$$

Node	u°	v°	θ
1	0.000	0.000	0.000
2	1.000	−0.246	−0.234
3	0.998	0.247	0.229
4	0.000	0.000	0.000

$$n_2 = 66.91 \text{ Hz}$$

Node	u°	v°	θ
1	0.000	0.000	0.000
2	0.194	−0.046	1.000
3	0.198	0.044	−0.392
4	0.000	0.000	0.000

Chapter 7

Vibrations of Space Trusses

This program, which is given in Appendix 4, is intended for determining the natural frequencies of vibration of three-dimensional structures whose members consist of uniform section rods joined together with smooth frictionless pins.

7.1 Description of Program

The rods can be of different sizes and of different elastic properties, and they are described by nodal points at their ends. Each node has three degrees of freedom ($u°$, $v°$ and $w°$), making a total of six degrees of freedom per element, as shown in Fig. 7.1.

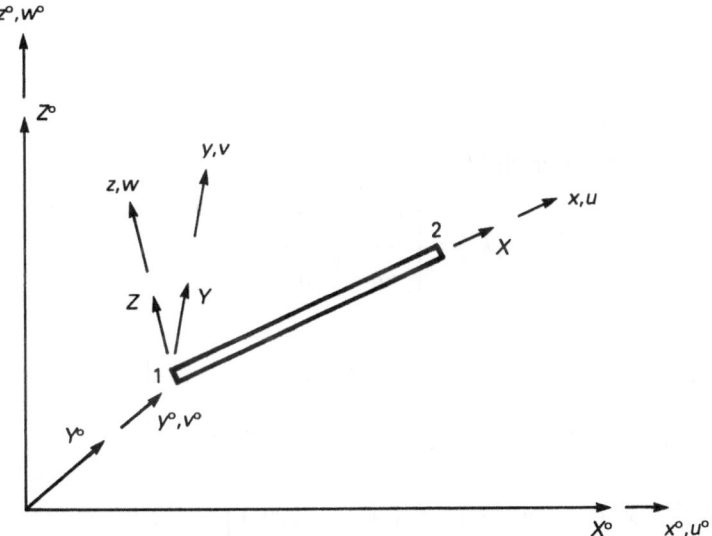

Fig. 7.1. Rod element in three-dimensional coordinate systems.

The elements are assumed to possess only axial stiffness and the stiffness matrix in global coordinates $[k^\circ]$ and the mass matrix in global coordinates $[m^\circ]$ are given in equations (7.1) and (7.2).

$$
[k^\circ] = \frac{AE}{l}
\begin{array}{ccc}
u_1^\circ & v_1^\circ & w_1^\circ
\end{array}
$$

$$
[k^\circ] = \frac{AE}{l}
\left[
\begin{array}{ccc|ccc}
C^2x,x^\circ & & & & & \\
Cx,x^\circ,Cx,y^\circ & C^2x,y^\circ & & & & \\
Cx,x^\circ Cx,z^\circ & Cx,y^\circ Cx,z^\circ & C^2x,z^\circ & & & \\
\hline
-C^2x,x^\circ & -Cx,x^\circ Cx,y^\circ & -Cx,x^\circ Cx,z^\circ & & & \\
-Cx,x^\circ Cx,y^\circ & -C^2x,y^\circ & -Cx,y^\circ Cx,z^\circ & & & \\
-Cx,x^\circ Cx,z^\circ & -Cx,y^\circ Cx,z^\circ & -C^2x,z^\circ & & &
\end{array}
\right.
$$

$$
\begin{array}{ccc}
u_2^\circ & v_2^\circ & w_2^\circ
\end{array}
$$

$$
\left.
\begin{array}{ccc}
& \text{Symmetrical} & \\
& & \\
\hline
C^2x,x^\circ & & \\
Cx,x^\circ Cx,y^\circ & C^2x,\ y^\circ & \\
Cx,x^\circ Cx,z^\circ & Cx,y^\circ Cx,z^\circ & C^2x,z^\circ
\end{array}
\right]
\begin{array}{c}
u_1^\circ \\
v_1^\circ \\
w_1^\circ \\
u_2^\circ \\
v_2^\circ \\
w_2^\circ
\end{array}
$$

$$(7.1)$$

which can be written in the form:

$$
[k^\circ] =
\left[
\begin{array}{c|c}
a & -a \\
\hline
-a & a
\end{array}
\right]
$$

where

$$
[a] = \frac{AE}{l}
\left[
\begin{array}{ccc}
C^2x,\ x^\circ & \text{Symmetrical} & \\
Cx,\ x^\circ Cx,\ y^\circ & C^2x,\ y^\circ & \\
Cx,\ x^\circ Cx,\ z^\circ & Cx,\ y^\circ Cx,z^\circ & C^2x,\ z^\circ
\end{array}
\right]
$$

$$
\left.
\begin{array}{l}
l = \sqrt{[(x_j^\circ - x_i^\circ)^2 + (y_j^\circ - y_i^\circ)^2 + (z_j^\circ - z_i^\circ)^2]} \\
Cx,\ x^\circ = (x_j^\circ - x_i^\circ)/l \\
Cx,\ y^\circ = (y_j^\circ - y_i^\circ)/l \\
Cx,\ z^\circ = (z_j^\circ - z_i^\circ)/l
\end{array}
\right\}.
$$

$$[m°] = \frac{\rho Al}{6} \begin{bmatrix} 2 & & & & & \\ 0 & 2 & & \text{symmetrical} & & \\ 0 & 0 & 2 & & & \\ 1 & 0 & 0 & 2 & & \\ 0 & 1 & 0 & 0 & 2 & \\ 0 & 0 & 1 & 0 & 0 & 2 \end{bmatrix} \begin{matrix} u_1° \\ v_1° \\ w_1° \\ u_2° \\ v_2° \\ w_2° \end{matrix}$$

$$\begin{matrix} u_1° & v_1° & w_1° & u_2° & v_2° & w_2° \end{matrix}$$

(7.2)

where,

$$A = \text{cross-sectional area}$$

$$l = \text{elemental length}$$

$$E = \text{Young's modulus}$$

$$\rho = \text{density}$$

7.2 Data

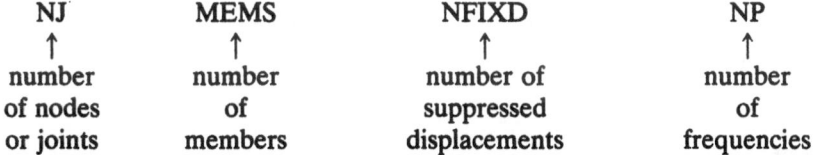

NJ	MEMS	NFIXD	NP
↑	↑	↑	↑
number of nodes or joints	number of members	number of suppressed displacements	number of frequencies

Nodal Coordinates

$$i = 1(1)NJ$$

$$x_i° \quad y_i° \quad z_i°$$

REPEAT

Member Details

$$i = 1(1)MEMS$$

I_i	J_i	A_i	E_i	RHO_i
nodes describing each element		cross-sectional area	elastic modulus	density

REPEAT

Details of Zero Displacements

$i = 1(1)\text{NFIXD}$

NPOSS_i = the node with the zero displacements

$\text{NXSUP}_i = 1$ if the displacement in the $x°$ direction is zero;
  else $\text{NXSUP}_i = 0$

$\text{NYSUP}_i = 1$ if the displacement in the $y°$ direction is zero;
  else $\text{NYSUP}_i = 0$

$\text{NZSUP}_i = 1$ if the displacement in the $z°$ direction is zero;
  else $\text{NZSUP}_i = 0$

→ REPEAT

NCONC = number of concentrated masses – if NCONC = 0 ignore (a)

(a) $i = 1(1)\text{NCONC}$

NPOS_i	XMCONC_i
nodal position of concentrated mass	magnitude of concentrated mass

→ REPEAT

Output

The output file should appear on a file called OUT9.OUT, and should include the following:

Details of the input file MAIN9DAT.DAT, plus,

$i = 1(1)\text{NP}$

$n_i = \sqrt{(1/\lambda_i)}/(2\pi)$ – frequencies

$\left\{ \;\; \right\}_i$ – eigenmode corresponding to "n_i"

→ REPEAT

7.3 Pin-jointed Space Truss Problems

Example 7.1

Determine the lowest natural frequency of vibration for the pin-jointed tripod shown in Fig. 7.2.

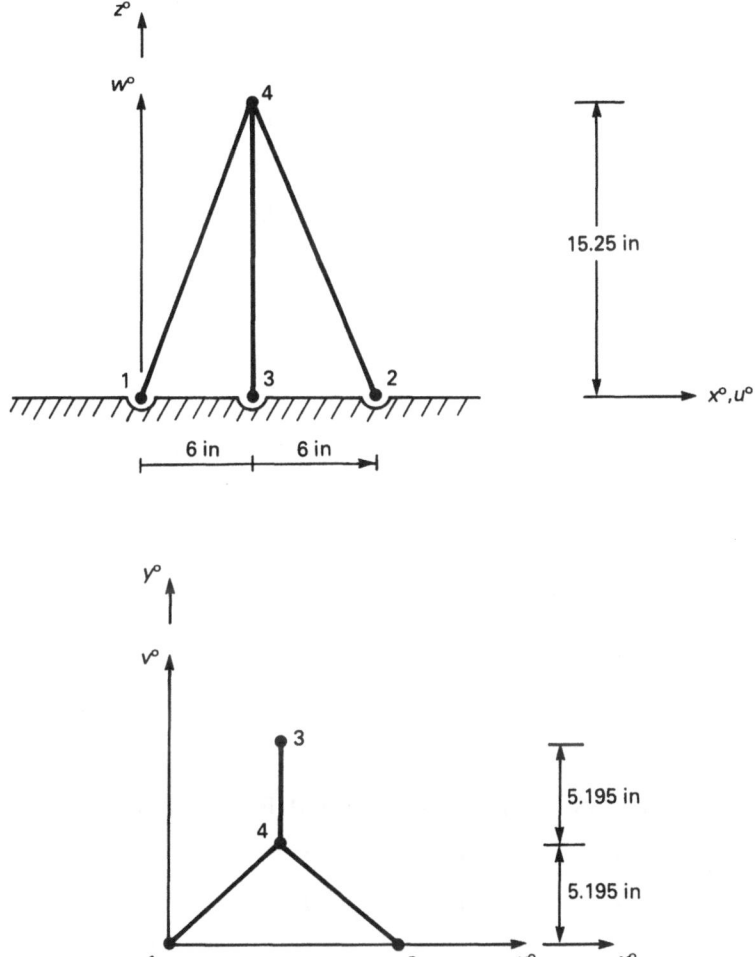

Fig. 7.2.   Pin-jointed tripod.

The following particulars apply to the members of the tripod:

Cross-sectional area $= 0.012$ in^2

Elastic modulus $= 30 \times 10^6$ lbf/in^2

Density $= 7.35 \times 10^{-4}$ lbf.s^2/in^4

Data

NJ = 4   NFIXD = 9   MEMS = 3   NP = 1

Nodal Coordinates

x°	y°	z°
0.00	0.00	0.00
12.00	0.00	0.00
6.00	10.39	0.00
6.00	5.195	15.25

Member Details

i	j	A	E	RHO
1	4	0.012	30×10^6	7.35×10^{-4}
2	4	0.012	30×10^6	7.35×10^{-4}
3	4	0.012	30×10^6	7.35×10^{-4}

Details of Zero Displacements

NPOSS	NXSUP	NYSUP	NZSUP
1	1	1	1
2	1	1	1
3	1	1	1

NCONC = 0

Results

1st eigenvalue $= 2.91 \times 10^{-8}$

1st frequency $= 933$ Hz

Node	u°	v°	w°	
1	0.0	0.0	0.0	
2	0.0	0.0	0.0	1st eigenmode – see Fig. 7.3
3	0.0	0.0	0.0	
4	1.0	0.712	−0.056	

Example 7.2

Determine the lowest natural frequency of vibration when the truss of Example 7.1 carries a mass of 0.00518 lbf.s^2/in at its free node. See Appendices A4.2 and A4.3.

Data

The data should be similar to those of Example 7.1, except that

$$\text{NCONC} = 1, \text{ plus the following}$$

$$4 \qquad 0.00518$$

Results

$$n_1 = 155.7 \text{ Hz}$$

Node	u°	v°	w°
1	0.000	0.000	0.000
2	0.000	0.000	0.000
3	0.000	0.000	0.000
4	1.000	0.712	−0.056

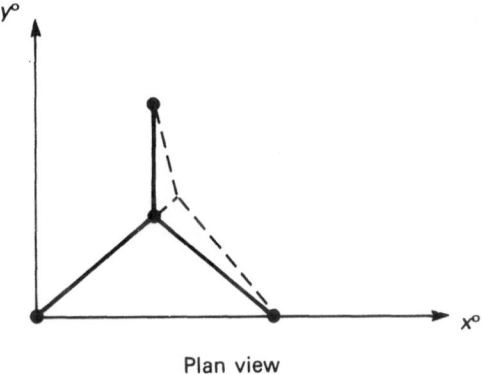

Plan view

Fig. 7.3. First eigenmode.

Chapter 8

Vibrations of Rigid-jointed Space Frames

This chapter describes how to use a FORTRAN computer program for determining resonant frequencies and eigenmodes of rigid-jointed space frames.

8.1 Description of Program

This program, which is given in Appendix 5, is intended for determining the natural frequencies of vibration of rigid-jointed three-dimensional structures. These structures are assumed to consist of one-dimensional members rigidly joined together. Each element has end nodes and there are six degrees of freedom per node ($u°$, $v°$, $w°$, $\theta_x°$, $\theta_y°$ and $\theta_z°$), making a total of twelve degrees of freedom per element, as shown in Fig. 8.1.

It is evident that with six degrees of freedom per node, any moderate or large size structure will require considerable storage space and to counteract this effect, the continuous reduction technique of Irons [9], is adopted.

This technique assumes that the displacements consist of "master" and "slave" displacements, so that when the reduction is complete, only master displacements remain to describe the vibration characteristics. The process, therefore, is to gradually build up the structural stiffness and mass matrices, and when a "slave" displacement is complete, it can be eliminated so long as lower "slaves" are eliminated. The resulting space saved by this technique can now be used to store "masters" and higher "slaves", until the latter are eliminated when complete.

Care must be taken in deciding which displacements should be described as "masters" and which as "slaves", as elimination of key slave displacements can cause the loss of some of the lower modes of vibration, but this is largely a matter of experience.

The members can have different sectional and material properties, and if there is a concentrated mass at a node, this can be added to the appropriate free master displacements.

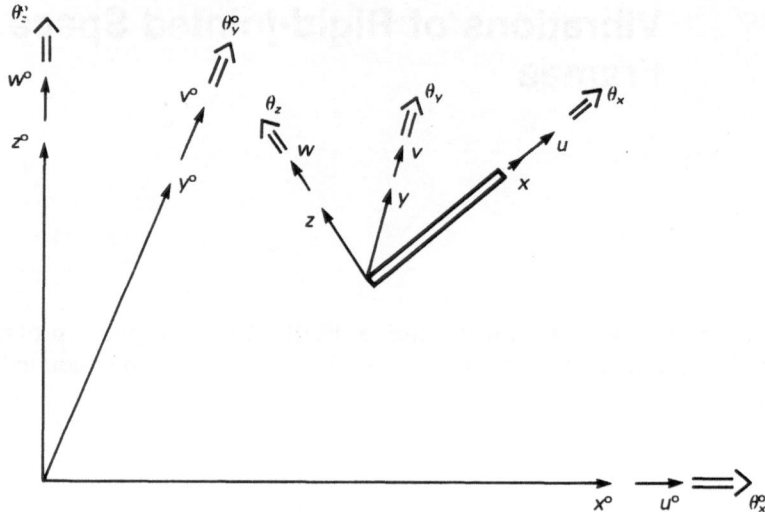

Fig. 8.1.   General one-dimensional element.

The stiffness matrix for the element $[k°]$ includes the effects of axial and torsional stiffness and also bending stiffness about the two principal axes of bending. This matrix, together with the mass matrix $[m°]$ is shown in equations (8.1) and (8.2):

$$[k°] = [DC]^T[k][DC] \tag{8.1}$$

$$[m°] = [DC]^T[m][DC] \tag{8.2}$$

where $[k]$ is shown in Table 8.1 and $[m]$ in Table 8.2. Where,

A = cross-sectional area

I_y = 2nd moment of area of the cross-section about the $x-z$ plane

I_z = 2nd moment of area of the cross-section about the $x-y$ plane

J = torsional constant of cross-section (see [10])

I_p = polar 2nd moment of area of cross-section

$L = l$ = elemental length

E = Young's modulus

ρ = density

$[DC]$ = a matrix of directional cosines [7]

Table 8.1. Elemental stiffness matrix $[k]$ for the general case of the one-dimensional member

	u_i	v_i	w_i	θ_{xi}	θ_{yi}	θ_{zi}	u_j	v_j	w_j	θ_{xj}	θ_{yj}	θ_{zj}	
$\dfrac{AE}{L}$													u_i
0	$\dfrac{12EI_z}{L^3}$												v_i
0	0	$\dfrac{12EI_y}{L^3}$											w_i
0	0	0	$\dfrac{GJ}{L}$										θ_{xi}
0	0	$\dfrac{-6EI_y}{L^2}$	0	$\dfrac{4EI_y}{L}$				Symmetrical					θ_{yi}
0	$\dfrac{6EI_z}{L^2}$	0	0	0	$\dfrac{4EI_z}{L}$								θ_{zi}
$\dfrac{-AE}{L}$	0	0	0	0	0	$\dfrac{AE}{L}$							u_j
0	$\dfrac{-12EI_z}{L^3}$	0	0	0	$\dfrac{-6EI_z}{L^2}$	0	$\dfrac{12EI_z}{L^3}$						v_j
0	0	$\dfrac{-12EI_y}{L^3}$	0	$\dfrac{6EI_y}{L^2}$	0	0	0	$\dfrac{12EI_y}{L^3}$					w_j
0	0	0	$\dfrac{-GJ}{L}$	0	0	0	0	0	$\dfrac{GJ}{L}$				θ_{xj}
0	0	$\dfrac{-6EI_y}{L^2}$	0	$\dfrac{2EI_y}{L}$	0	0	0	$\dfrac{6EI_y}{L^2}$	0	$\dfrac{4EI_y}{L}$			θ_{yj}
0	$\dfrac{6EI_z}{L^2}$	0	0	0	$\dfrac{2EI_z}{L}$	0	$\dfrac{-6EI_z}{L^2}$	0	0	0	$\dfrac{4EI_z}{L}$		θ_{zj}

Table 8.2. Elemental mass matrix $[m]$ for the generalized case of the one-dimensional member

	u_1	v_1	w_1	θ_{x1}	θ_{y1}	θ_{z1}	u_2	v_2	w_2	θ_{x2}	θ_{y2}	θ_{z2}	
	$\dfrac{1}{3}$												u_1
	0	$\dfrac{13}{15}+\dfrac{6I_z}{5AL^2}$											v_1
	0	0	$\dfrac{13}{15}+\dfrac{6I_y}{5AL^2}$					Symmetrical					w_1
	0	0	0	$\dfrac{I_p}{3A}$									θ_{x1}
	0	0	$\dfrac{-11L}{210}-\dfrac{I_y}{10AL}$	0	$\dfrac{L^2}{105}+\dfrac{2I_y}{15A}$								θ_{y1}
	0	$\dfrac{11L}{210}+\dfrac{I_z}{10AL}$	0	0	0	$\dfrac{L^2}{105}+\dfrac{2I_z}{15A}$							θ_{z1}
	$\dfrac{1}{6}$	0	0	0	0	0	$\dfrac{1}{3}$						u_2
	0	$\dfrac{9}{70}-\dfrac{6I_z}{5AL^2}$	0	0	0	$\dfrac{13L}{420}-\dfrac{I_z}{10AL}$	0	$\dfrac{13}{35}+\dfrac{6I_z}{5AL^2}$					v_2
	0	0	$\dfrac{9}{70}-\dfrac{6I_y}{5AL^2}$	0	$\dfrac{-13L}{420}+\dfrac{I_y}{10AL}$	0	0	0	$\dfrac{13}{35}+\dfrac{6I_y}{5AL^2}$				w_2
	0	0	0	$\dfrac{I_p}{6A}$	0	0	0	0	0	$\dfrac{I_p}{3A}$			θ_{x2}
	0	0	$\dfrac{13L}{420}-\dfrac{I_y}{10AL}$	0	$\dfrac{L^2}{140}-\dfrac{I_y}{30A}$	0	0	0	$\dfrac{11L}{210}+\dfrac{I_y}{10AL}$	0	$\dfrac{L^2}{105}+\dfrac{2I_y}{15A}$		θ_{y2}
	0	$\dfrac{13L}{420}+\dfrac{I_z}{10AL}$	0	0	0	$\dfrac{-L^2}{140}-\dfrac{I_z}{30A}$	0	$\dfrac{-11L}{210}-\dfrac{I_z}{10AL}$	0	0	0	$\dfrac{L^2}{105}+\dfrac{2I_z}{15A}$	θ_{z2}

$$[DC] = \begin{bmatrix} \zeta & & & \\ & \zeta & & \\ \hline & & \zeta & \\ \hline & & & \zeta \end{bmatrix}$$

$$[\zeta] = \begin{bmatrix} Cx, x° & Cx, y° & Cx, z° \\ Cy, x° & Cy, y° & Cy, z° \\ Cz, x° & Cz, y° & Cz, y° \end{bmatrix}$$

$$Cx, x° = (x_2° - x_1°)/l$$
$$Cx, y° = (y_2° - y_1°)/l$$
$$Cx, z° = (z_2° - z_1°)/l$$
$$l = [(x_2° - x_1°)^2 + (y_2° - y_1°)^2 + (z_2° - z_1°)^2]$$

The other directional cosines Cy, $x°$, Cy, $y°$, etc., can be obtained by constructing a triangular plane $(x-y)$ through the centroid of the section and parallel with a principal plane of bending, as shown in Figs. 8.2 and 8.3 and reference [7].

The principal $x-y$ bending plane of an element is defined by the triangular plane i, j, k, as shown in Fig. 8.2. The principal $x-z$ bending plane of the element is orthogonal to its $x-y$ bending plane and is determined in the usual manner [7]. If it is convenient, the kth node is defined as an imaginary node

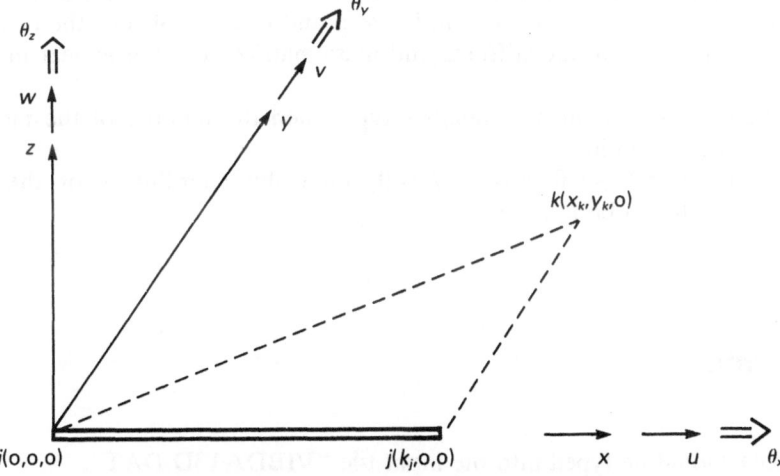

Fig. 8.2.   Element in local coordinates $x-y-z$.

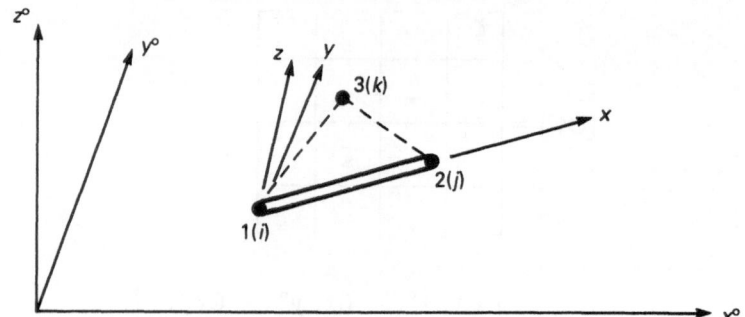

Fig. 8.3.   Triangular plane in local and global coordinates.

at any convenient point. Imaginary nodes do not increase the size of the stiffness and mass matrices.

It should be noted that for many problems, the node k can be an existing node on the structure, but this depends on the directions of the principal axes of bending of the element's cross-section.

If a node is to be completely eliminated, then the nodal number is followed by three zeroes, e.g. if the node to be eliminated is 19, then in place of 19, type 19000, when this node appears for the last time. If all the displacements are to be eliminated, except for the $u°$, $v°$ and $w°$ displacements at a node, then the nodal number should be made negative, e.g. if only $u_{19}°$, $v_{19}°$ and $w_{19}°$ are required displacements at node 19, type -19, when this node appears for the last time.

It should be noted that it is a simple matter to alter the program so that only one displacement remains or displacements apart from $u°$, $v°$ and $w°$ are left.

If a node is completely fixed it should be defined as a zero node. For such a node, all its "displacements" will be zero and because of this, the elements corresponding to it in the stiffness and mass matrices will not appear in these matrices.

If there is more than one member type, then the number of the member types should be fed in.

If $i = 0$ or/and $j = 0$ or/and $k = 0$, then the coordinates of the zero node/nodes should be fed in order.

8.2 Data

The data should be typed into the input file "VIBDAT3D.DAT",

NJ = Number of non-zero joints.

NF = Number of fixed (or zero) nodes.

NSUP = Number of zero displacements at pin-jointed supports, etc.

NIMK = Number of imaginary nodes. (These are required in addition to the actual nodes to define the oxy plane of an element – see Figs. 8.2 and 8.3).

N = Number of free displacements remaining. (These are the displacements that are left after the reduction of the slave displacements and the elimination of the displacements corresponding to a zero nodal point).

M1 = Number of frequencies.

NMATL = Number of member types.

MEMS = Number of members or elements.

Nodal Coordinates

   ┌─ $i = 1$ To NJ + NIMK

   │   Input x_i°, y_i° and z_i°

   └→ REPEAT

Member Types

   ┌─ $i = 1$ To NMATL

   │ RHO_i = density

   │ E_i = elastic modulus

   │ G_i = rigidity modulus

   │ CSA_i = cross-sectional area

   │ TC_i = torsional constant

   │ $SMAY_i$ = 2nd moment of area about local x–y plane) (see Fig. 8.3)

   │ $SMAZ_i$ = 2nd moment about local x–z plane

   │ $POLAR_i$ = polar 2nd moment of area

   └→ REPEAT

Element Topology

$i = 1$ To MEMS

Input i, j and k nodes for each element.

(The node k is used to define one of the principal axes of bending of the element, so that $i-j-k$ is the oxy plane of the element. The 2nd moment of area I_y (SMAY), corresponds to the oxy plane – see Fig. 8.3.)

If NMATL > 1, then type in the member type for this member.

→ REPEAT

Pinned Nodes, etc.

$i = 1$ To NSUP

NS$_i$ – Type in the finished displacement position of the zero displacement. This displacement must be a u, v or w displacement and it should be ensured that this node is not completely eliminated.

→ REPEAT

Concentrated Masses

NCONC = number of concentrated masses

$i = 1$ To NCONC

POS$_i$ = "u" displacement position of the mass – the nodes must not be completely eliminated and the mass is added to the mass matrix in the $u°$, $v°$ and $w°$ directions at the particular node.

MASS$_i$ = Value of mass at POS$_i$

→ REPEAT

Zero Nodes' Coordinates

Type in the $x°$, $y°$ and $z°$ coordinates of the zero nodes, as appropriate and in order. It should be noted that the numerical precision of the eigenvalues has been set to 0.001 (or 0.1%), and this can be increased or decreased, but care should be taken to ensure that the precision is not so small that it exceeds the precision of the machine.

The results appear in the output file "OUT11.OUT", as follows:

Details of input data, plus,

$$1\text{st eigenvalue} \quad = \lambda_1$$

$$1\text{st frequency} \quad = n_1$$

$$\{\ \}_1 \quad 1\text{st eigenmode}$$

$$2\text{nd eigenvalue} = \lambda_2$$

$$2\text{nd frequency} \quad = n_2$$

$$\{\ \}_2 \quad 2\text{nd eigenmode}$$

$$3\text{rd eigenvalue} = \lambda_3$$

$$3\text{rd frequency} \quad = n_3$$

$$\{\ \}_3 \quad 3\text{rd eigenmode}$$

etc.

N.B. The eigenmode output is with reference only to the displacements remaining after reduction.

8.3 Rigid-jointed Space Frame Problems

Example 8.1

Determine the three lowest natural frequencies of vibration for the rigid-jointed space frame of Fig. 8.4. It may be assumed that the frame is pinned at

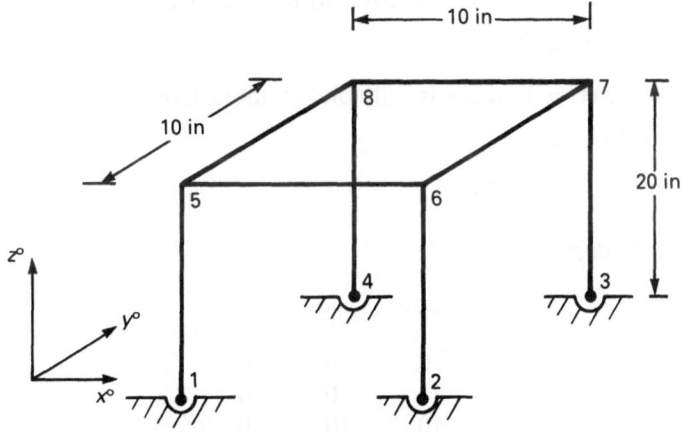

Fig. 8.4.   Rigid-jointed space frame.

its base and that it is composed of symmetrical section members with the following properties:

$$\rho = 7.35 \times 10^{-4} \text{ lbf s}^2/\text{in}^4$$

$$E = 30 \times 10^6 \text{ lbf/in}^2$$

$$G = 1.15 \times 10^7 \text{ lbf/in}^2$$

$$\text{CSA} = 6.3 \times 10^{-2} \text{ in}^2$$

$$\text{TC} = 1.3 \times 10^{-3} \text{ in}^4$$

$$\text{SMAY} = 3.26 \times 10^{-4} \text{ in}^4$$

$$\text{SMAZ} = 3.26 \times 10^{-4} \text{ in}^4$$

$$\text{POLAR} = 6.51 \times 10^{-4} \text{ in}^4$$

As the structure is symmetrical, it will be necessary to make the structure slightly unsymmetrical, by altering the coordinates of some of the FREE NODES. This process is necessary to determine frequencies and eigenmodes, apart from the first.

Data

$$\text{NJ} = 8$$

$$\text{NF} = 0$$

$$\text{NSUP} = 12 \text{ (the displacements } u_1{}^\circ, v_1{}^\circ, w_1{}^\circ, u_2{}^\circ \rightarrow w_4{}^\circ = 0)$$

$$\text{NIMK} = 0 \text{ (as the sections are symmetrical)}$$

$$N = 24 \text{ (these correspond to } u_1{}^\circ, v_1{}^\circ, w_1{}^\circ, u_2{}^\circ, v_2{}^\circ,$$

$$w_2{}^\circ, u_3{}^\circ, v_3{}^\circ, w_3{}^\circ \text{ to } u_8{}^\circ, v_8{}^\circ, w_8{}^\circ)$$

$$\text{M1} = 3$$

$$\text{NMATL} = 1 \text{ (there is only one member type)}$$

$$\text{MEMS} = 8$$

Nodal Coordinates

$x_i{}^\circ$	$y_i{}^\circ$	$z_i{}^\circ$
0	0	0
10	0	0
10	10	0
0	10	0

$$
\begin{array}{ccc}
1E-2 & 1E-2 & 20.01 \\
10.01 & 1E-2 & 19.99 \\
10.01 & 10.01 & 20.1 \\
1E-2 & 10.01 & 19.99
\end{array}
$$

Material Properties

RHO	= 7.35E−4	E	= 3E7	G	= 1.15E7
CSA	= 6.3E−2	TC	= 1.3E−3	SMAY	= 3.26E−4
SMAZ	= 3.26E−4	POLAR	= 6.51E−4		

Element Topology

$$
\begin{array}{rrr}
i & j & k \\
-1 & 5 & 8 \\
-2 & 6 & 7 \\
-3 & 7 & 6 \\
-4 & 8 & 5 \\
5 & 6 & 2 \\
-5 & 8 & 7 \\
-6 & 7 & 8 \\
-7 & -8 & 5
\end{array}
$$

Zero Displacements
at pin-jointed supports, etc.

$$
\begin{array}{rrrl}
1 & 2 & 3 & (@\ \text{node } 1) \\
4 & 5 & 6 & (@\ \text{node } 2) \\
7 & 8 & 9 & (@\ \text{node } 3) \\
10 & 11 & 12 & (@\ \text{node } 4)
\end{array}
$$

$$\text{NCONC} = 0$$

Results

$$\lambda_1 = 3.007E-4$$
$$n_1 = 9.18 \text{ Hz}$$

$$\lambda_2 = 2.998E-4$$
$$n_2 = 9.19 \text{ Hz}$$

$$\lambda_3 = 2.348E-4$$
$$n_3 = 10.39 \text{ Hz}$$

Example 8.2

Determine the three lowest natural frequencies of vibration for the rigid-jointed space frame of Example 8.1, assuming that it is fixed at its base. The data should be fed in as follows:

$$NJ \quad = 4$$

$$NF \quad = 4$$

$$NSUP \quad = 0$$

$$NIMK \quad = 0$$

$$N \quad = 12 \text{ (i.e. } u_1^\circ, v_1^\circ, w_1^\circ, u_2^\circ \to u_4^\circ, v_4^\circ, w_4^\circ)$$

$$M1 \quad = 3$$

$$NMATL \quad = 2$$

$$MEMS \quad = 8$$

x_i°	y_i°	z_i°
1E-2	1E-2	20.01
10.01	1E-2	19.99
10.01	10.01	20.01
1E-2	10.01	19.99

RHO	7.35E-4	E	= 3E7	G	= 1.15E7
CSA	= 6.3E-2	TC	= 1.3E-3	SMAY	= 3.26E-4
SMAZ	= 3.26E-4	POLAR=	6.51E-4		

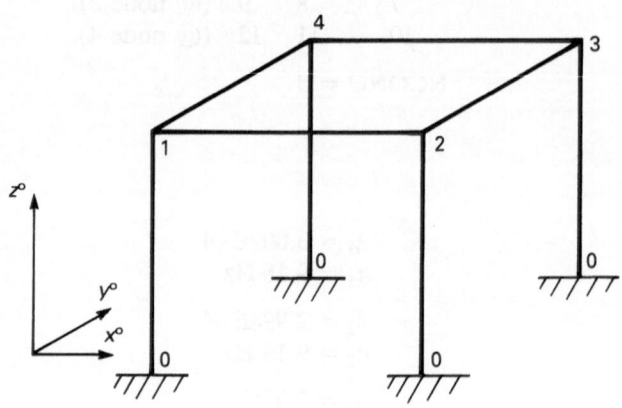

Fig. 8.5.

$$
\begin{array}{ccc}
i & j & k \\
0 & 1 & 4 \\
0 & 2 & 3 \\
0 & 3 & 2 \\
0 & 4 & 1 \\
1 & 2 & 3 \\
-1 & 4 & 3 \\
-2 & 3 & 4 \\
-3 & -4 & 1
\end{array}
$$

NCONC = 0

Coordinates for Zero Nodes

$$
\begin{array}{ccc}
x^\circ & y^\circ & z^\circ \\
0 & 0 & 0 \quad \text{(Element 1)} \\
10 & 0 & 0 \quad \text{(Element 2)} \\
10 & 10 & 0 \quad \text{(Element 3)} \\
0 & 10 & 0 \quad \text{(Element 4)}
\end{array}
$$

Results

$$\lambda_1 = 6.478\text{E}{-5}$$
$$n_1 = 19.77 \text{ Hz}$$

$$\lambda_2 = 6.475\text{E}{-5}$$
$$n_2 = 19.78 \text{ Hz}$$

$$\lambda_3 = 2.354\text{E}{-5}$$
$$n_3 = 32.81 \text{ Hz}$$

Example 8.3

Determine the three lowest natural frequencies of vibration for the rigid-jointed space frame of Example 8.2, assuming that there are additional concentrated masses of 5×10^{-4} lbf s^2/in and 1×10^{-3} lbf s^2/in at nodes 1 and 3 respectively and that the sectional properties of the vertical members have twice the magnitude of the horizontal members. See Appendices A5.2 and A5.3.

Data

$$
\begin{aligned}
\text{NJ} &= 4 \\
\text{NF} &= 4 \\
\text{NSUP} &= 0 \\
\text{NIMK} &= 0 \\
\text{N} &= 12 \\
\text{M1} &= 3 \\
\text{NMATL} &= 2 \\
\text{MEMS} &= 8
\end{aligned}
$$

$x_i{}^\circ$	$y_i{}^\circ$	$z_i{}^\circ$
1E–2	1E–2	20.01
10.01	1E–2	19.99
10.01	10.01	20.01
1E–2	10.01	19.99

Details of Member Types

RHO	= 7.35E–4	E	= 3E7	G	= 1.15E7
CSA	= 6.3E–2	TC	= 1.3E–3	SMAY	= 3.26E–4
SMAZ	= 3.26E–4	POLAR	= 6.51E–4		
RHO	= 7.35E–4	E	= 3E7	G	= 1.15E7
CSA	= 0.126	TC	= 2.6E–3	SMAY	= 6.52E–4
SMAZ	= 6.52E–4	POLAR	= 1.302E–3		

i	j	k	Member Type
0	1	4	2
0	2	3	2
0	3	2	2
0	4	1	2
1	2	3	1
–1	4	3	1
–2	3	4	1
–3	–4	1	1

NCONC = 2

POS$_1$ = 1 ($u°$ displacement position for 1st concentrated mass)

MASS$_1$ = 5E–4

POS$_2$ = 7 ($u°$ displacement position for 2nd concentrated mass)

MASS$_2$ = 1E–3

Coordinates for Zero Nodes

$x°$	$y°$	$z°$
0	0	0
10	0	0
10	10	0
0	10	0

Results

$$\lambda_1 = 6.902\text{E}{-}5$$
$$n_1 = 19.16 \text{ Hz}$$

$$\lambda_2 = 6.930\text{E}{-}5$$
$$n_2 = 19.12 \text{ Hz}$$

$$\lambda_3 = 2.591\text{E}{-}5$$
$$n_3 = 31.27 \text{ Hz}$$

Example 8.4

Determine the three lowest natural frequencies of vibration for the rigid-jointed space frame shown in Figs. 8.6 and 8.7. The particulars of the frame are given in Appendix 5 (A5.4). As the elements are of circular cross-section, they have an infinite number of principal axes of bending, hence the kth node can be chosen from any nodal point other than the current "i" and "j" nodes of the particular element of interest.

The first three eigenmodes of the tower are given in Fig. 8.8 in plan view. See Appendices A5.4 and A5.5. From Fig. 8.8, it can be seen that the first two eigenmodes, which are flexural, are effectively of the same type and occur because of symmetry of the frame. The third eigenmode is a lozenging type.

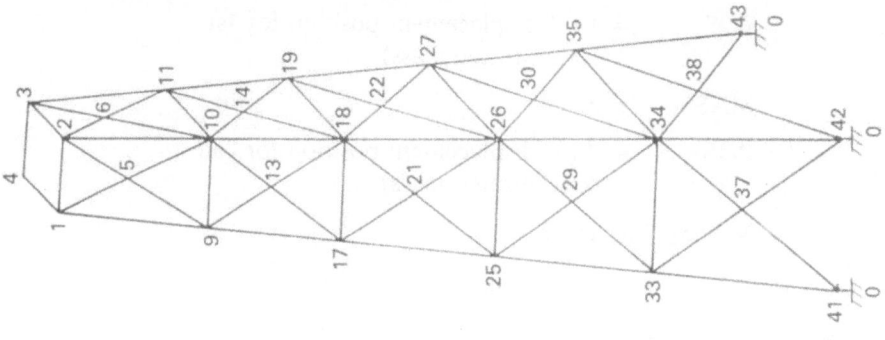

Fig. 8.7.   Mathematical model of tower.

Fig. 8.6.   Model tower.

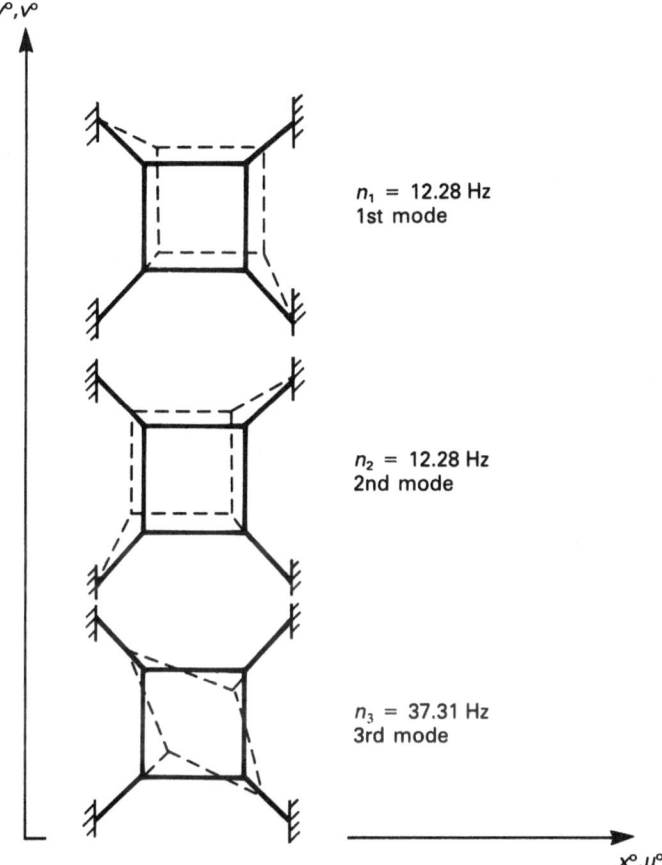

Fig. 8.8.   Eigenmodes of model tower in plan view.

N.B. It should be noted that, owing to the massive use of reduction, it is possible that local modes may have been missed out. It should also be noted that in the above analysis, the eigenmodes of the structure were described by u°, v° and w° displacements only.

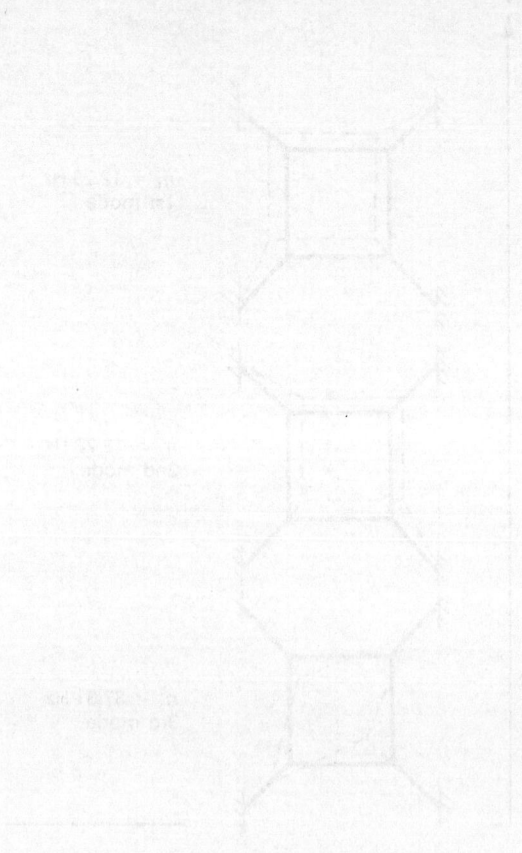

Fig. X.X. Elementaries of model drawn in plan view.

N.B. It should be noted that, owing to the massive use of reinforcement it is possible that local modes may have been missed out. It should also be noted that in the above analysis, the eigenmodes of the structure were described by x, y and z displacements.

Chapter 9

Vibration of Grillages

This chapter describes how to use a FORTRAN computer program which can calculate the resonant frequencies and eigenmodes of orthogonal and skew grids.

9.1 Description of Program

This program can analyse orthogonal or skew grids, as shown in Fig. 9.1. The grids can be simply-supported or fixed or can have any combination of simply-supported and fixed supports, etc.

The element in local coordinates is shown in Fig 9.2, and its stiffness matrix in local coordinates is shown by equation (9.1). This matrix can be seen to be a combination of a beam element and a torque bar.

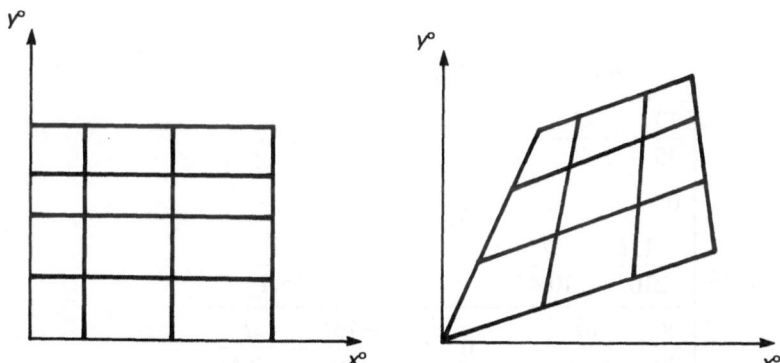

Fig. 9.1. Orthogonal and skew grids.

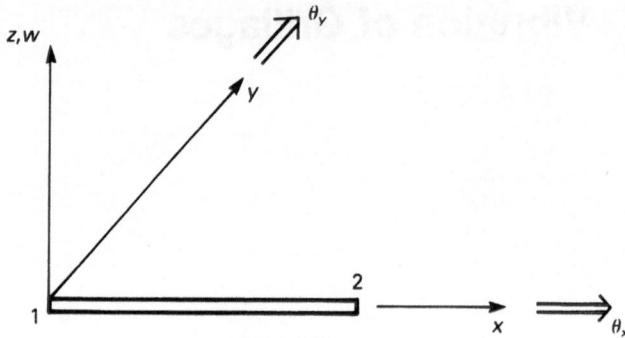

Fig. 9.2.   Beam element for grillage.

$$[k] = \begin{array}{cc} \begin{array}{cccccc} w_1 & \theta_{x1} & \theta_{y1} & w_2 & \theta_{x2} & \theta_{y2} \end{array} \\ \left[\begin{array}{cccccc} 12EI/l^3 & & & & & \\ 0 & GJ/l & & & & \\ -6EI/l^2 & 0 & 4EI/l & & & \\ -12EI/l^3 & 0 & 6EI/l^2 & 12EI/l^3 & & \\ 0 & -GJ/l & 0 & 0 & GJ/l & \\ -6EI/l^2 & 0 & 2EI/l & 6EI/l^2 & 0 & 4EI/l \end{array}\right] \begin{array}{c} w_1 \\ \theta_{x1} \\ \theta_{y1} \\ w_2 \\ \theta_{x2} \\ \theta_{y2} \end{array} \end{array}$$

$$(9.1)$$

The elemental mass matrix is given by equation (9.2), which can also be seen to be a combination of beam and a torque bar element.

$$[m] = \rho AL \begin{array}{c} \begin{array}{ccc} w_1 \qquad\quad & \theta_{x1} \quad & \theta_{y1} \end{array} \\ \left[\begin{array}{ccc} \dfrac{13}{35} + \dfrac{6I}{5Al^2} & & \\[2ex] 0 & \dfrac{I_p}{3A} & \\[2ex] \dfrac{-11l}{210} - \dfrac{I}{10Al} & 0 & \dfrac{l^2}{105} + \dfrac{2I}{15A} \\[2ex] \dfrac{9}{70} - \dfrac{6I}{5Al^2} & 0 & \dfrac{-13l}{420} + \dfrac{I}{10Al} \\[2ex] 0 & \dfrac{I_p}{6A} & 0 \\[2ex] \dfrac{13l}{420} - \dfrac{I}{10Al} & 0 & \dfrac{l^2}{140} - \dfrac{I}{30A} \end{array}\right] \end{array}$$

$$
\begin{array}{ccc}
w_2 & \theta_{x2} & \theta_{y2}
\end{array}
$$

$$
\left.\begin{array}{ccc}
& & \\
& & \\
& & \\
\hline
\dfrac{13}{35} + \dfrac{6I}{5Al^2} & & \\
0 & \dfrac{I_p}{3A} & \\
\dfrac{11l}{210} + \dfrac{I}{10AL} & 0 & \dfrac{l^2}{105} + \dfrac{2I}{15A}
\end{array}\right]
\begin{array}{l}
w_1 \\
\theta_{x1} \\
\theta_{y1} \\
w_2 \\
\theta_{x2} \\
\theta_{y2}
\end{array}
\qquad (9.2)
$$

where

l = length of element

I = 2nd moment of area about the $x-y$ plane

J = torsional constant

I_p = 2nd polar moment of area about the "x" axis

A = cross-sectional area

E = elastic modulus

G = rigidity modulus

ρ = density.

In practice, however, it is necessary to obtain the elemental stiffness and mass matrices in the global coordinates of Fig. 9.3. This is because some of the grillage members point in directions other than that of Fig. 9.2.

Transformation of the stiffness and mass matrices of equations (9.1) and (9.2) into global axes is achieved as follows:

$$[k^\circ] = \text{elemental stiffness matrix in global axes}$$

$$= [DC]^T [k] [DC]$$

$$[m^\circ] = \text{elemental mass matrix in global axes}$$

$$= [DC]^T [m] [DC]$$

where,

$$
[DC] = \left[\begin{array}{c|c}
\zeta & O_3 \\
\hline
O_3 & \zeta
\end{array}\right]
$$

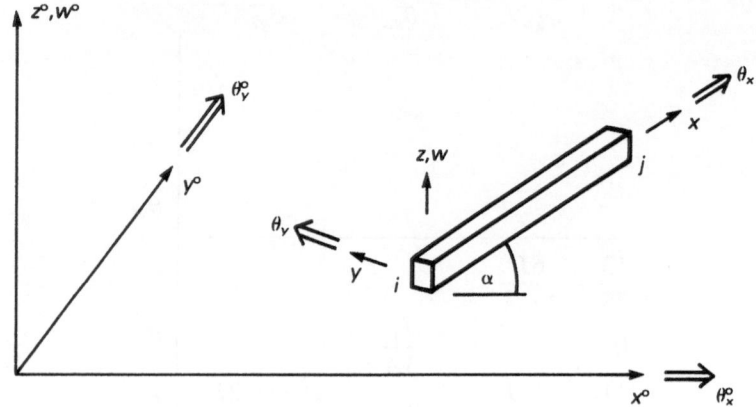

Fig. 9.3. Grillage element in global coordinates.

$$[\zeta] = \begin{bmatrix} 1 & 0 & 0 \\ 0 & c & s \\ 0 & -s & c \end{bmatrix}$$

O_3 = a null matrix of order 3.

$c = \cos \alpha$

$s = \sin \alpha$

α is defined in Fig. 9.3

9.2 Data

The data should be fed into the file as follows:

NJ   = Number of nodes.

NFIXD = Number of nodes with zero displacements.

NP   = Number of eigenvalues.

MEMX = Number of "x" direction members.

MEMY = Number of "y" direction members.

I   = 1(1)NJ

X(I)   = x coordinate of the Ith node.

Y(I)   = y coordinate of the Ith node.

REPEAT

E = Young's modulus.

G = Rigidity of shear modulus.

RHO = material density

CSAX = cross-sectional area of "x direction" members

CSAY = cross-sectional area of "y direction" members

SMAX = 2nd moment of area of "x direction" members

SMAY = 2nd moment of area of "y direction" members

TCX = torsional constant of "x direction" members

TCY = torsional constant of "y direction" members

POLX = polar 2nd moment of area of "x direction" members

POLY = polar 2nd moment of area of "y direction" members

Element Topology

┌─ MEM = 1 To MEMS

│ Feed "i" node followed by "j" node

└→ REPEAT

Details of Zero Displacements

┌─ $i = 1$ To NFIXD

│ $NPOSS_i$ = node number of node with zero displacements

│ If w displacement is zero type 1; else type 0

│ If θ_x displacement is zero type 1; else type 0

│ If θ_y displacement is zero type 1; else type 0

└→ REPEAT

Concentrated Masses

NCONC = Number of concentrated masses.

┌─ $i = 1$ To NCONC

│ $NPOS_i$ – nodal position of concentrated mass

│ $XMCONC_i$ – value of concentrated mass at the above position

└→ REPEAT

9.3 Grillage Problems

Example 9.1

Determine the lowest natural frequencies for the grillage of Fig. 9.4 [11], which is simply-supported at nodes 1, 2, 3, 4, 5 and 6 (Fig. 9.5).
The material properties, etc., are as follows:

$$\text{RHO} = 7860 \text{ kg/m}^3$$
$$E = 2 \times 10^{11} \text{ N/m}^2$$
$$G = 7.69 \times 10^{10} \text{ N/m}^2$$
$$\text{CSA} = 0.004 \text{ m}^2$$
$$\text{SMA} = 1.25 \times 10^{-5} \text{ m}^4$$
$$\text{TC} = \text{POL} = 2.5 \times 10^{-5} \text{ m}^4$$

Data

$$\text{NJ} \quad = 12$$
$$\text{NFIXD} = 6$$
$$\text{NP} \quad = 3$$
$$\text{MEMX} = 9$$
$$\text{MEMY} = 4$$

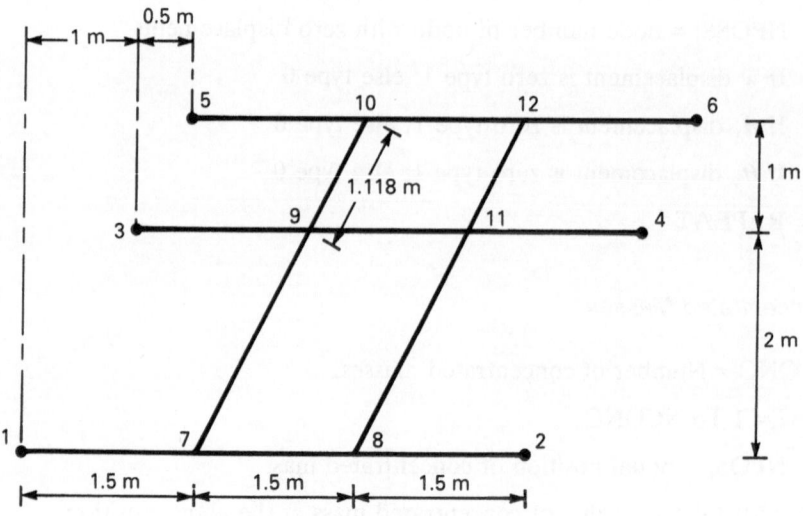

Fig. 9.4.   Skew grillage.

x_i	y_i
0	0
4.5	0
1	2
5.5	2
1.5	3
6	3
1.5	0
3	0
2.5	2
3	3
4	2
4.5	3

E   = 2E11

G   = 7.69E10

RHO   = 7860

CSAX = 4E−3

CSAY = 4E−3

SMAX = 1.25E−3

SMAY = 1.25E−3

TCX   = 2.5E−5

TCY   = 2.5E−5

POLX = 2.5E−5

POLY = 2.5E−5

Element Topology

i	j
1	7
7	8
8	2
3	9
9	11
11	4
5	10
10	12
12	6
7	9

8	11
9	10
11	12

Zero Displacement Details

1	1	0	0
2	1	0	0
3	1	0	0
4	1	0	0
5	1	0	0
6	1	0	0

Concentrated Masses

$$\text{NCONC} = 0$$

The results are as follows:

$$\lambda_1 = 8.9143\text{E}-5$$

$$n_1 = 16.857 \text{ Hz}$$

$$\lambda_2 = 6.1797\text{E}-5$$

$$n_2 = 20.246 \text{ Hz}$$

$$\lambda_3 = 8.8892\text{E}-6$$

$$n_3 = 53.381 \text{ Hz}$$

Example 9.2

Determine the three lowest natural frequencies of vibration for the grid of Fig. 9.4, assuming that there are additional concentrated masses of 50 kg and 40 kg at nodes 9 and 11, respectively. See Appendices A6.2 and A6.3.

Data

12	6	3
9	4	

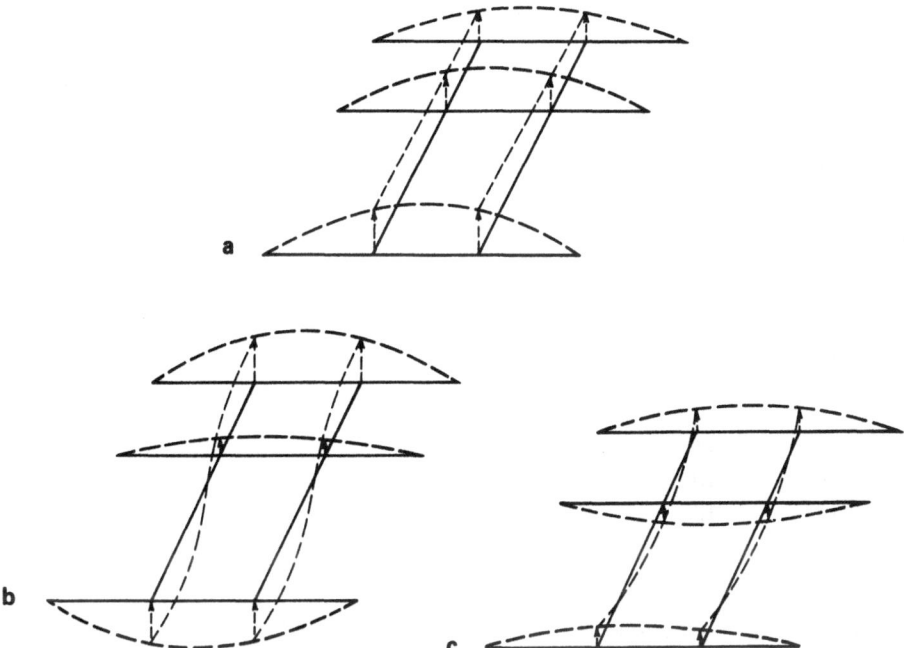

Fig. 9.5.   Vibration modes for skew grillage. **a** Mode 1 (16.85 Hz). **b** Mode 2 (20.25 Hz). **c** Mode 3 (53.4 Hz).

Nodal Coordinates

x_i	y_i
0	0
4.5	0
1	2
5.5	2
1.5	3
6	3
1.5	0
3	0
2.5	2
3	3
4	2
4.5	3

2E11   7.69E10   7860

4E−3   4E−3   1.25E−5   1.25E−5   2.5E−5

2.5E−5   2.5E−5   2.5E−5

Element Topology

i	j
1	7
7	8
8	2
3	9
9	11
11	4
5	10
10	12
12	6
7	9
8	11
9	10
11	12

Suppressed Displacements

1	1	0	0
2	1	0	0
3	1	0	0
4	1	0	0
5	1	0	0
6	1	0	0

Concentrated Masses

$$\text{NCONC} = 2$$

$$\text{POS}_1 = 9 \qquad \text{MASS}_1 = 50$$

$$\text{POS}_2 = 11 \qquad \text{MASS}_2 = 40$$

The results are as follows:

$$\lambda_1 = 1.0576\text{E}-4$$

$$n_1 = 15.48 \text{ Hz}$$

$$\lambda_2 = 6.3945\text{E}-5$$

$$n_2 = 19.90 \text{ Hz}$$

$$\lambda_3 = 1.0408\text{E}-5$$

$$n_3 = 49.33 \text{ Hz}$$

Example 9.3

Determine the three lowest natural frequencies of vibration for the grid of Fig. 9.6, which is fixed at its left-most and right-most supports. The material and other properties of its members are as in Example 9.1.

Data

As for Example 9.1, except for suppressed displacement positions, which are as follows:

Details of Zero Displacements

1	1	1	1
2	1	1	1
3	1	1	1
4	1	1	1
5	1	1	1
6	1	1	1

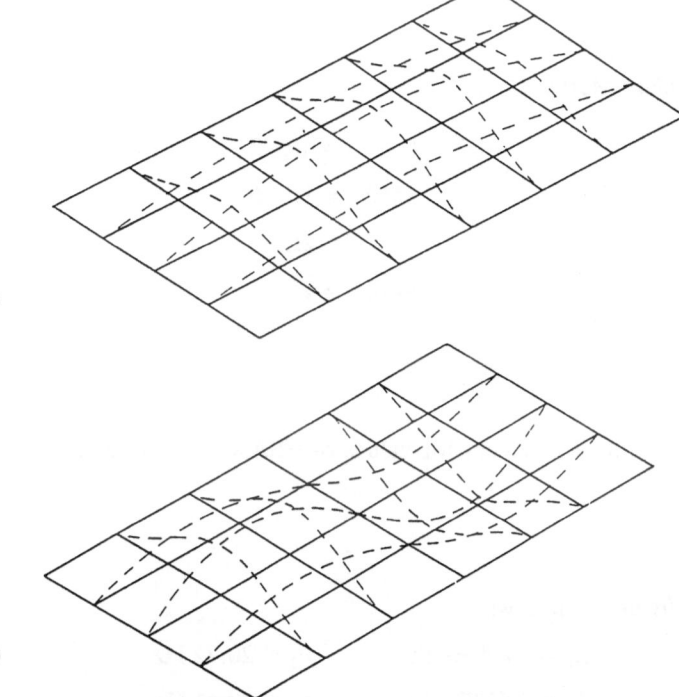

a

b

Fig. 9.6. Eigenmodes for clamped-supported grillage. **a** First mode. **b** Second mode.

The results are as follows:

$$\lambda_1 = 1.768E-5; \qquad n_1 = 37.84 \text{ Hz}$$
$$\lambda_2 = 1.130E-5; \qquad n_2 = 47.34 \text{ Hz}$$
$$\lambda_3 = 3.913E-6; \qquad n_3 = 80.46 \text{ Hz}$$

Example 9.4

Determine the three lowest natural frequencies of vibration for the grid of Fig. 9.4, assuming that the x and y direction members have the following sectional properties:

"x" Direction Members

$$\text{CSA} = 5E-3$$
$$\text{SMA} = 2E-5$$
$$\text{TC} = 3E-5$$
$$\text{POL} = 3E-5$$

"y" Direction Members

$$\text{CSA} = 3E-3$$
$$\text{SMA} = 1E-5$$
$$\text{TC} = 2E-5$$
$$\text{POL} = 2E-5$$

Data

As for Example 9.1, except for values of CSAX, CSAY, etc.

Output

The *results* are as follows:

$$\lambda_1 = 5.899E-5; \qquad n_1 = 20.72 \text{ Hz}$$
$$\lambda_2 = 4.623E-5; \qquad n_2 = 23.41 \text{ Hz}$$
$$\lambda_3 = 1.089E-5; \qquad n_3 = 48.24 \text{ Hz}$$

Example 9.5

Determine the three lowest natural frequencies and the corresponding eigen-modes for the grillage of Example 2.5 of reference [11].

Results

The first three natural frequencies of vibration are:

$$n_1 = 100.7 \text{ Hz}$$

$$n_2 = 102.6 \text{ Hz}$$

$$n_3 = 110.4 \text{ Hz}$$

and the normalised eigenmodes corresponding to n_1 and n_2 are shown in Fig. 9.6.

From Fig. 9.6, it can be seen that the first eigenmode is symmetrical and consists of one half-wave in both directions. The second eigenmode is unsymmetrical, and consists of one half-wave in the $x°$ direction and two half-waves in the $y°$ direction.

Example 9.6

Determine the three lowest natural frequencies of vibrations, together with their eigenmodes for the ship grillage of reference [12].

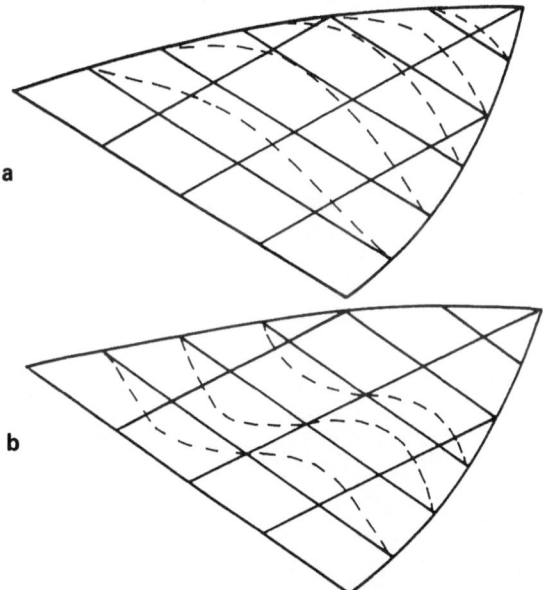

Fig. 9.7. Eigenmodes for ship grillage. **a** First mode. **b** Second mode.

Results

The three lowest natural frequencies of vibration are:

$$n_1 = 13.59 \text{ Hz}$$

$$n_2 = 27.5 \text{ Hz}$$

$$n_3 = 33.15 \text{ Hz}$$

and the two eigenmodes corresponding to n_1 and n_2 are shown in Fig. 9.7.

From Fig. 9.7, it can be seen that the first eigenmode is symmetrical and consists of one half-wave in both directions. The second eigenmode is unsymmetrical, and consists of two half-waves in the $x°$ direction and one half-wave in the $y°$ direction.

Appendix 1. Computer Program for the Vibration of Plane Pin-jointed Trusses

A 1.1 The Computer Program

```
C     ******************************************
C     *VIBRATION OF PLANE PIN JOINTED TRUSSES*
C     ******************************************
C
      DIMENSION STIFF(50,50),SMAS(50,50),
     1 EV(10),VEC(50,10),XX(50,51)
C     MODIFIED FOR IBM PC ON 21st JULY 1990
C     COPYRIGHT OF DR.C.T.F.ROSS.
      OPEN(5,FILE='MAIN6DAT.DAT',STATUS='OLD')
      OPEN(6,FILE='OUT6.OUT',STATUS='OLD')
      READ(5,*) NN,LEMS,NFIXD,NV
      WRITE(6,100) NN,LEMS,NFIXD,NV
  100 FORMAT(6X, 23H ***BASIC PARAMETERS***//
     1 5X, 27H NUMBER OF NODES . . . . . ,I2/
     1 5X, 27H NUMBER OF MEMBERS . . . . ,I2/
     1 5X, 43H NUMBER OF NODES WITH ZERO DISPLACEMENTS . ,I2/
     1 5X, 27H NUMBER OF EIGEN VALUES. . ,I2)
      NN2=NN*2
      NP1=NN2+1
      CALL XMAIN(STIFF,SMAS,NN,LEMS,NFIXD,NV,NN2,
     1 NP1,EV,VEC,XX)
      CLOSE(6)
      CLOSE(5)
      STOP
      END
C
      SUBROUTINE XMAIN(STIFF,SMAS,NN,LEMS,NFIXD,NV,NN2,
     1 NP1,EV,VEC,XX)
      DIMENSION STIFF(NN2,NN2),SMAS(NN2,NN2),
     1 EV(NV),VEC(NN2,NV),XX(NN2,NP1),
     4 NSUP(50),NPOSS(50),NXSUP(50),NYSUP(50),X(50),Y(50),
     5 ST(4,4),SM(4,4),
     6 LL(50),M(50),BB(50)
      NJOINTS=NN
      NNODES=2
      NDF=2
C
      CALL NULL(STIFF,NN2,NN2)
      CALL NULL(SMAS,NN2,NN2)
```

```
      WRITE(6,110)
 110  FORMAT(///30H NODES WITH ZERO DISPLACEMENTS)
      READ(5,*) (NPOSS(II),NXSUP(II),NYSUP(II),II=1,NFIXD)
      WRITE(6,101)  (NPOSS(II),NXSUP(II),NYSUP(II),II=1,NFIXD)
 101  FORMAT(3I5)
      WRITE(6,111)
      DO 777 II=1,50
      NSUP(II)=0
 777  CONTINUE
      DO 77 II=1,NFIXD
      IF(NXSUP(II).NE.0)NSUP(II*2-1)=NPOSS(II)*2-1
      IF(NYSUP(II).NE.0)NSUP(II*2)=NPOSS(II)*2
 77   CONTINUE
 111  FORMAT(///23H ***NODAL POINT DATA***/
     1 2X, 4HNODE, 6X, 1HX, 9X, 1HY)
      DO 2 II=1,NN
      READ(5,*) X(II),Y(II)
      WRITE(6,102) II,X(II),Y(II)
 2    CONTINUE
 102  FORMAT(I5,2F10.3)
C
C
      WRITE(6,112)
 112  FORMAT(/ //6X, 18H***ELEMENT DATA***/
     1 4X, 1HI, 7X, 1HJ, 7X, 4HAREA, 12X, 1HE, 14X, 3HRHO)
      DO 3 LEM=1,LEMS
      WRITE(*,778)LEM
 778  FORMAT(' ELEMENT NO.',I5,' UNDER COMPUTATION')
      READ(5,*) I, J, A, E, RHO
      WRITE(6,103) I, J, A, E, RHO
 103  FORMAT(I5,3X,I5,3E15.5)
      XL=SQRT((X(J)-X(I))**2+(Y(J)-Y(I))**2)
      C=(X(J)-X(I))/XL
      S=(Y(J)-Y(I))/XL
      ST(1,1)=C**2
      ST(1,2)=C*S
      ST(1,3)=-C**2
      ST(1,4)=-C*S
      ST(2,1)=C*S
      ST(2,2)=S**2
      ST(2,3)=-C*S
      ST(2,4)=-S**2
      ST(3,1)=-C**2
      ST(3,2)=-C*S
      ST(3,3)=C**2
      ST(3,4)=C*S
```

```
      ST(4,1)=-C*S
      ST(4,2)=-S**2
      ST(4,3)=C*S
      ST(4,4)=S**2
      CN=A*E/XL
      CALL SCPROD(ST,4,4,CN)
C
      CALL NULL(SM,4,4)
      SM(1,1)=2.0
      SM(2,2)=2.0
      SM(3,3)=2.0
      SM(4,4)=2.0
      SM(1,3)=1.0
      SM(3,1)=1.0
      SM(2,4)=1.0
      SM(4,2)=1.0
      CN=RHO*A*XL/6.0
      CALL SCPROD(SM,4,4,CN)
C
      CALL ASSEMB(I,J,4,ST,STIFF,NN2)
      CALL ASSEMB(I,J,4,SM,SMAS,NN2)
  3   CONTINUE
C
C
      IF(NCONC.EQ.0)GOTO 121
      DO 122 II=1,NCONC
      READ(5,*) NPOS,XMCONC
      WRITE(6,123) NPOS,XMCONC
 123  FORMAT(I5,1X,E12.4)
      I1=2*NPOS-1
      I9=I1+1
      SMAS(I1,I1)=SMAS(I1,I1)+XMCONC
      SMAS(I9,I9)=SMAS(I9,I9)+XMCONC
 122  CONTINUE
 121  CONTINUE
C
      NFIXD2=NFIXD*NDF
      CALL SUPPRESS(STIFF,NSUP,NN2,NFIXD2)
      WRITE(*,999)
 999  FORMAT('  THE STIFFNESS MATRIX IS BEING INVERTED')
      CALL INVMX(STIFF,NN2,0.0,LL,M,BB)
      CALL PRODAB(STIFF,SMAS,NN2)
      WRITE(*,779)
 779  FORMAT('  THE EIGENVALUES ARE BEING DETERMINED')
      D=0.001
      CALL EIGEN(STIFF,NN2,EV,VEC,NV,D,XX,NP1)
```

```
        CALL EIGPRIN2(VEC,EV,NJOINTS,NDF,NN2,NV)
        RETURN
        END
C
        SUBROUTINE ASSEMB(I,J,NDF2,ST,A,N)
        DIMENSION A(N,N),ST(NDF2,NDF2)
        NDF=NDF2/2
        I1=NDF*I-NDF
        J1=NDF*J-NDF
        DO 10 II=1,2
        DO 10 JJ=1,2
        MM=I1+II
        MN=I1+JJ
        NM=J1+II
        N9=J1+JJ
        A(MM,MN)=A(MM,MN)+ST(II,JJ)
        A(NM,MN)=A(NM,MN)+ST(II+NDF,JJ)
        A(MM,N9)=A(MM,N9)+ST(II,JJ+NDF)
        A(NM,N9)=A(NM,N9)+ST(II+NDF,JJ+NDF)
  10    CONTINUE
        RETURN
        END
C
        SUBROUTINE SUPPRESS(A,NSUP,N,NFIXD)
        DIMENSION A(N,N),NSUP(1)
        DO 10 II=1,NFIXD
        NS=NSUP(II)
        IF(NS.NE.0)A(NS,NS)=A(NS,NS)*1.0E12+1.0E12
  10    CONTINUE
        RETURN
        END
C
        SUBROUTINE VALEIG(A,Y,X2,N,D,N1)
        DIMENSION A(1),
     2  Y(200),Z(200)
C       MODIFIED 20-11-78
        X1=100000.0
        DO 1 I=1,N
        Y(I)=1.0
  1     CONTINUE
  2     X2=-100000.0
        DO 3 I=1,N
        IF (X2.LT.Y(I))      X2=Y(I)
  3     CONTINUE
        X=1.0/X2
        DO 4 I=1,N
```

```
       SUM=0.0
               DO  5   J=1,N
       SUM=SUM+A((J-1)*N1+I)*Y(J)
5              CONTINUE
       Z(I)=SUM*X
4      CONTINUE
       IF(ABS((X1-X2)/X2).GT.D) GO TO  6
       GO TO 66
 6     DO 7 I=1,N
       Y(I)=Z(I)
7      CONTINUE
       X1=X2
         GO TO  2
66     DO 8 I=1,N
       IF(ABS(Z(I)).LE.0.000000001) GOTO 8
       IF(ABS((Y(I)-Z(I))/Z(I)).LE.D) GOTO 8
14              DO 10 J=1,N
                Y(J)=Z(J)
10              CONTINUE
                X1=X2
                GO TO  2
8      CONTINUE
       X = 0.0
       DO 11 I=1,N
       IF (ABS(X).LT.ABS(Y(I))) X=Y(I)
11     CONTINUE
       DO 12 I=1,N
       Y(I)=Y(I)/X
12     CONTINUE
       RETURN
       END
C
       SUBROUTINE EIGEN(A,N,AM,VEC,M1,D,XX,NP1)
       DIMENSION A(1),
      1 VEC(1),AM(1),XX(1),
      2 X(200),IG(200)
         MN=N
       NN = N
       CALL VALEIG(A,X,XM,NN,D,N)
         M=1
         DO 2 I=1,NN
       I3=(M-1)*N+I
       VEC(I3)=X(I)
       XX(I3)=VEC(I3)
2        CONTINUE
       AM(M) = XM
```

```
       DO 20 M=2,M1
       DO 4 I=1,NN
       IF(ABS(XX((M-2)*N+I)-1.0).LT.0.00001)IR=I
4      CONTINUE
       IG(M-1)=IR
       DO 5 I=1,NN
       XX((MN-M+2)*N+MN-I+1)=A((I-1)*N+IR)
5      CONTINUE
       DO 6 I=1,NN
       DO 6 J=1,NN
       I3=(J-1)*N+I
       I4=(M-2)*N+I
       I5=(MN-M+2)*N+MN-J+1
       A(I3)=A(I3)-XX(I4)*XX(I5)
6      CONTINUE
       DO 14 I=1,NN
       IF(I.EQ.IR) GO TO 14
       IF(I.GT.IR) K1=I-1
       IF(I.LE.IR) K1=I
             DO  7 J=1,NN
             IF(J.EQ.IR) GO TO 7
             IF(J.GT.IR) K2=J-1
       IF(J.LE.IR) K2=J
       A((K2-1)*N+K1)=A((J-1)*N+I)
7            CONTINUE
14     CONTINUE
       NN = NN-1
       M3=NN
       IF (M.NE.MN)  GO TO  11
       XM=A(1)
       X(1)=1.0
       GO TO 12
11     CALL VALEIG(A,X,XM,NN,D,N)
12     DO 13 I=1,NN
       XX((M-1)*N+I)=X(I)
13     CONTINUE
       AM(M)=XM
       M11 = M-1
       MM1=1000-M11
       DO 20 M8=MM1,999
       M12 = M3+1
       M2=1000-M8
       M22 =IG(M2)+1
       MM22=1000-M22
       MM12=1000-M12
       DO 15 I3=MM12,MM22
```

```
          I=1000-I3
          X(I)=X(I-1)
  15      CONTINUE
          J=IG(M2)
          X(J)=0.0
          SUM=0.0
          DO 16 I= 1,M12
         SUM=SUM+XX((MN-M2+1)*N+MN-I+1)*X(I)
  16      CONTINUE
          XK=(AM(M2)-XM)/SUM
          DO 17 I= 1,M12
          X(I)=XX((M2-1)*N+I)-XK*X(I)
  17      CONTINUE
          SUM=0.0
          DO 18 I=1,M12
          IF (ABS(SUM).LT.ABS(X(I)))SUM=X(I)
  18      CONTINUE
          DO 19 I=1,M12
          X(I)=X(I)/SUM
  19      CONTINUE
          M3=M3+1
          IF(M2.NE.1)GO TO 20
          DO 3 I=1,M3
          VEC((M-1)*N+I)=X(I)
   3      CONTINUE
  20      CONTINUE
          RETURN
          END
C
          SUBROUTINE  EIGPRIN2(VEC,XLAM,NJOINTS,NDOF,N,NV)
C
C         PRINTS RESULTS OF EIGENVALUE ANALYSIS:
C         FREQUENCY AND MODE SHAPE,INCORPORATING SUPPRESSED DOF'S.
C
          DIMENSION  VEC(N,NV),XLAM(NV)
          WRITE(6,30)
  30      FORMAT(///12X, 20H***EIGEN SOLUTION***)
          PI=3.1415927
          DO 150 IEIG=1,NV
            PAR=1.0/(2.0*PI)*SQRT(1.0/XLAM(IEIG))
          WRITE(6,60) IEIG, PAR
  60      FORMAT(///7X, I3, 1X, 12H FREQUENCY =,E14.6/
         1 8X, 29H--------------------------)
            WRITE(6,80)
  80        FORMAT(//10X,11H MODE SHAPE/
         1 10X,6H  NODE,5X, 37H RELATIVE DISPLACEMENTS(GLOBAL COORD))
```

```
          DO 90 IPOIN=1,NJOINTS
      IP=IPOIN*NDOF-NDOF
        WRITE(6,120) IPOIN,(VEC(IP+IDOF,IEIG),IDOF=1,NDOF)
 90       CONTINUE
120   FORMAT(9X, I5, 3X, E10.3, 4X, E10.3, 4X, E10.3)
150   CONTINUE
      RETURN
      END
C
      SUBROUTINE PRODAB(A,B,N)
      DIMENSION A(N,N),B(N,N),VEC(100,1)
C     (A)=(A)*(B)
      DO 1 I=1,N
      DO 2 J=1,N
      VEC(J,1)=0.0
      DO 2 K=1,N
      VEC(J,1)=VEC(J,1)+A(I,K)*B(K,J)
 2    CONTINUE
      DO 1 J=1,N
      A(I,J)=VEC(J,1)
 1    CONTINUE
      RETURN
      END
C
      SUBROUTINE INVMX(A,N,D,L,M,B)
      DIMENSION A(1),B(1),L(1),M(1)
      I5=0
 3    I1=N-1
      DO 1 I=1,I1
      I2=I+1
      DO 2 J=I2,N
      I3=(I-1)*N+J
      I4=(J-1)*N+I
      B(J)=A(I3)
      A(I3)=A(I4)
 2    A(I4)=B(J)
 1    CONTINUE
      IF (I5.EQ.1) RETURN
      NK=-N
      DO 80 K=1,N
      NK=NK+N
      L(K)=K
      M(K)=K
      KK=NK+K
      BIGA=A(KK)
      DO 20 J=K,N
```

```
        IZ=N*(J-1)
        DO 20 I=K,N
        IJ=IZ+I
10      IF (ABS(BIGA)-ABS(A(IJ))) 15,20,20
15      BIGA=A(IJ)
        L(K)=I
        M(K)=J
20      CONTINUE
        J=L(K)
        IF (J-K) 35,35,25
25      KI=K-N
        DO 30 I=1,N
        KI=KI+N
        HOLD=-A(KI)
        JI=KI-K+J
        A(KI)=A(JI)
30      A(JI)=HOLD
35      I=M(K)
        IF (I-K) 45,45,38
38      JP=N*(I-1)
        DO 40 J=1,N
        JK=NK+J
        JI=JP+J
        HOLD=-A(JK)
        A(JK)=A(JI)
40      A(JI)=HOLD
45      IF (BIGA) 48,46,48
46      D=0.0
        RETURN
48      DO 55 I=1,N
        IF (I-K) 50,55,50
50      IK=NK+I
        A(IK)=A(IK)/(-BIGA)
55      CONTINUE
        DO 65 I=1,N
        IK=NK+I
        HOLD=A(IK)
        IJ=I-N
        DO 65 J=1,N
        IJ=IJ+N
        IF(I-K) 60,65,60
60      IF (J-K) 62,65,62
62      KJ=IJ-I+K
        A(IJ)=HOLD*A(KJ)+A(IJ)
65      CONTINUE
        KJ=K-N
```

```
        DO 75 J=1,N
        KJ=KJ+N
        IF (J-K) 70,75,70
70      A(KJ)=A(KJ)/BIGA
75      CONTINUE
        D=D*BIGA
        A(KK)=1.0/BIGA
80      CONTINUE
        K=N
100     K=(K-1)
        IF (K) 150,150,105
105     I=L(K)
        IF (I-K) 120,120,108
108     JQ=N*(K-1)
        JR=N*(I-1)
        DO 110 J=1,N
        JK=JQ+J
        HOLD=A(JK)
        JI=JR+J
        A(JK)=-A(JI)
110     A(JI)=HOLD
120     J=M(K)
        IF (J-K) 100,100,125
125     KI=K-N
        DO 130 I=1,N
        KI=KI+N
        HOLD=A(KI)
        JI=KI-K+J
        A(KI)=-A(JI)
130     A(JI)=HOLD
        GO TO 100
150     I5=1
        GO TO 3
        END
C
        SUBROUTINE SCPROD(A,M,N,CN)
        DIMENSION A(M,N)
C       (A)=CN*(A)
        DO 1 I=1,M
        DO 1 J=1,N
        A(I,J)=CN*A(I,J)
1       CONTINUE
        RETURN
        END
C
        SUBROUTINE NULL(A,M,N)
```

```
      DIMENSION A(M,N)
      DO 1 I=1,M
      DO 1 J=1,N
      A(I,J)=0.0
 1    CONTINUE
      RETURN
      END
```

A 1.2 A Typical Input File for MAIN6DAT.DAT (Example 4.1)

```
4   3   3   2
1   1   1
2   1   1
3   1   1
0.  0.
17.32  0.
23.09  0.
17.32  10.
1   4   0.06   3.E7   7.35E-4
2   4   0.06   3.E7   7.35E-4
3   4   0.06   3.E7   7.35E-4
0
```

A 1.3 A Typical Output File for OUT6.OUT (Example 4.1)

```
        ***BASIC PARAMETERS***

        NUMBER OF NODES . . . . .  4
        NUMBER OF MEMBERS . . . .  3
        NUMBER OF NODES WITH ZERO DISPLACEMENTS .  3
        NUMBER OF EIGEN VALUES. .  2



NODES WITH ZERO DISPLACEMENTS
     1     1     1
     2     1     1
     3     1     1
```

```
***NODAL POINT DATA***
  NODE      X          Y
    1     0.000      0.000
    2    17.320      0.000
    3    23.090      0.000
    4    17.320     10.000
```

```
    ***ELEMENT DATA***
   I      J        AREA              E               RHO
   1      4    0.60000E-01     0.30000E+08      0.73500E-03
   2      4    0.60000E-01     0.30000E+08      0.73500E-03
   3      4    0.60000E-01     0.30000E+08      0.73500E-03
```

```
        ***EIGEN SOLUTION***



    1   FREQUENCY =   0.206357E+04
        _____



        MODE SHAPE
          NODE        RELATIVE DISPLACEMENTS(GLOBAL COORD)
            1      0.144E-11        0.201E-11
            2      0.124E-07        0.142E-12
            3      0.114E-11       -0.428E-12
            4      0.100E+01        0.132E+00



    2   FREQUENCY =   0.366185E+04
        _____



        MODE SHAPE
          NODE        RELATIVE DISPLACEMENTS(GLOBAL COORD)
            1      0.294E-12        0.423E-11
            2     -0.512E-08        0.122E-11
            3     -0.202E-11        0.146E-11
            4     -0.132E+00        0.100E+01
```

Appendix 2. Computer Program for the Vibration of Continuous Beams

A 2.1 The Computer Program

```
C      ********************
C      *VIBRATION OF BEAMS*
C      ********************
C
       DIMENSION STIFF(60,60),SMAS(60,60),
      1 XX(60,61),EV(20),VEC(60,20)
         OPEN(5,FILE='MAIN7DAT.DAT',STATUS='OLD')
         OPEN(6,FILE='OUT7.OUT',STATUS='OLD')
         READ(5,*) LEMS,NFIXD,NV
  100  FORMAT(3I5)
       NN=2*LEMS+2
       NP1=NN+1
       CALL XMAIN(STIFF,SMAS,NN,NP1,LEMS,NFIXD,NV,EV,VEC,XX)
       CLOSE(5)
       CLOSE(6)
       STOP
       END
C
C
       SUBROUTINE XMAIN(STIFF,SMAS,NN,NP1,LEMS,NFIXD,NV,EV,VEC,XX)
       DIMENSION STIFF(NN,NN),SMAS(NN,NN),
      1 EV(NV),VEC(NN,NV),XX(NN,NP1),
      2 NSUP(50),NPOSS(50),NXSUP(50),NYSUP(50),ST(4,4),
      3 SM(4,4),
      2 LL(60),M(60),BB(60)
       REAL L
       NJOINTS=LEMS+1
       NNODES=2
       NDF=2
       CALL NULL(STIFF,NN,NN)
       CALL NULL(SMAS,NN,NN)
       WRITE(6,100) LEMS,NFIXD,NV
  100  FORMAT(23H ***BASIC PARAMETERS***//
      1 5X, 27H NUMBER OF ELEMENTS . . . . .,I2/
      2 5X, 47H NUMBER OF NODES WITH ZERO DISPLACEMENTS. . . .,I2/
      4 5X, 27H NUMBER OF EIGEN VALUES . .,I2)
        READ(5,*) (NPOSS(II),NXSUP(II),
      9 NYSUP(II),II=1,NFIXD)
       WRITE(6,110)
```

```
110  FORMAT(///41H DETAILS OF NODES WITH ZERO DISPLACEMENTS)
     WRITE(6,101) (NPOSS(II),NXSUP(II),
   9 NYSUP(II),II=1,NFIXD)
101  FORMAT(3I5)
     DO 77 II=1,NFIXD*2
     NSUP(II)=0
77   CONTINUE
     DO 777 II=1,NFIXD
     IF (NXSUP(II).NE.0) NSUP(II*2-1)=NPOSS(II)*2-1
     IF (NYSUP(II).NE.0) NSUP(II*2)=NPOSS(II)*2
777  CONTINUE
     READ(5,*) E,RHO
     WRITE(6,102) E,RHO
102  FORMAT(///5X, 16H YOUNGS MODULUS=,E15.6/
   1       5X, 16H DENSITY=          ,E15.6)
     WRITE(6,111)
111  FORMAT(///19H ***ELEMENT DATA***/
   1 4X, 1HI, 4X, 1HJ, 12X, 3HSMA,
   2 17X, 3HCSA, 15X, 6HLENGTH)
     DO 1 LEM=1,LEMS
     READ(5,*) I, J, SMA, CSA, L
     WRITE(6,103) I, J, SMA, CSA, L
103  FORMAT(2I5,3E20.8)
     ST(1,1)=12.0/L**3
     ST(1,2)=-6.0/L**2
     ST(1,4)=ST(1,2)
     ST(2,1)=ST(1,2)
     ST(4,1)=ST(1,2)
     ST(3,3)=ST(1,1)
     ST(1,3)=-12.0/L**3
     ST(3,1)=ST(1,3)
     ST(2,2)=4.0/L
     ST(4,4)=ST(2,2)
     ST(2,3)=6.0/L**2
     ST(3,2)=ST(2,3)
     ST(2,4)=2.0/L
     ST(4,2)=ST(2,4)
     ST(3,4)=6.0/L**2
     ST(4,3)=ST(3,4)
C
     CN=E*SMA
     CALL SCPROD(ST,4,4,CN)
C
     SM(1,1)=156.0
     SM(3,3)=156.0
     SM(1,2)=-22.0*L
```

```
        SM(2,1)=-22.0*L
        SM(1,3)=54.0
        SM(3,1)=54.0
        SM(1,4)=13.0*L
        SM(4,1)=13.0*L
        SM(2,2)=4.0*L*L
        SM(4,4)=4.0*L*L
        SM(2,3)=-13.0*L
        SM(3,2)=-13.0*L
        SM(2,4)=-3.0*L*L
        SM(4,2)=-3.0*L*L
        SM(3,4)=22.0*L
        SM(4,3)=22.0*L
C
        CN=RHO*CSA*L/420.0
        CALL SCPROD(SM,4,4,CN)
C
        CALL ASSEMB(I,J,4,ST,STIFF,NN)
        CALL ASSEMB(I,J,4,SM,SMAS,NN)
   1    CONTINUE
        READ(5,*) NCONC
        WRITE(6,120) NCONC
 120    FORMAT(//31H NUMBER OF CONCENTRATED MASSES=,I2)
        IF(NCONC.EQ.0) GOTO 3
        WRITE(6,122)
 122    FORMAT(//2X, 4HNODE, 6X, 8HCON MASS, 7X, 10HCON M.M.I.)
        DO 2 II=1,NCONC
        READ(5,*) NPOS, XMAS, XMMI
        WRITE(6,121) NPOS, XMAS, XMMI
 121    FORMAT(I5,4X,E12.4,4X,E12.4)
        I1=2*NPOS-1
        I9=I1+1
        SMAS(I1,I1)=SMAS(I1,I1)+XMAS
        SMAS(I9,I9)=SMAS(I9,I9)+XMMI
   2    CONTINUE
   3    CONTINUE
C
        NFIXD2=NFIXD*NDF
        CALL SUPPRESS(STIFF,NSUP,NN,NFIXD2)
        CALL INVMX(STIFF,NN,0.0,LL,M,BB)
        CALL PRODAB(STIFF,SMAS,NN)
        D=0.001
        CALL EIGEN(STIFF,NN,EV,VEC,NV,D,XX,NP1)
        CALL EIGPRIN2(VEC,EV,NJOINTS,NDF,NN,NV)
        RETURN
        END
```

```
C
      SUBROUTINE ASSEMB(I,J,NDF2,ST,A,N)
      DIMENSION A(N,N),ST(NDF2,NDF2)
      NDF=NDF2/2
      I1=NDF*I-NDF
      J1=NDF*J-NDF
      DO 10 II=1,2
      DO 10 JJ=1,2
      MM=I1+II
      MN=I1+JJ
      NM=J1+II
      N9=J1+JJ
      A(MM,MN)=A(MM,MN)+ST(II,JJ)
      A(NM,MN)=A(NM,MN)+ST(II+NDF,JJ)
      A(MM,N9)=A(MM,N9)+ST(II,JJ+NDF)
      A(NM,N9)=A(NM,N9)+ST(II+NDF,JJ+NDF)
  10  CONTINUE
      RETURN
      END
C
      SUBROUTINE SUPPRESS(A,NSUP,N,NFIXD)
      DIMENSION A(N,N),NSUP(1)
      DO 10 II=1,NFIXD
      NS=NSUP(II)
      IF(NSUP(II).NE.0)A(NS,NS)=A(NS,NS)*1.0E12
  10  CONTINUE
      RETURN
      END
C
      SUBROUTINE VALEIG(A,Y,X2,N,D,N1)
      DIMENSION A(1),
     2 Y(200),Z(200)
C     MODIFIED 20-11-78
      X1=100000.0
      DO 1 I=1,N
      Y(I)=1.0
   1  CONTINUE
   2  X2=-100000.0
      DO 3 I=1,N
      IF (X2.LT.Y(I))     X2=Y(I)
   3  CONTINUE
      X=1.0/X2
      DO 4 I=1,N
      SUM=0.0
            DO  5  J=1,N
      SUM=SUM+A((J-1)*N1+I)*Y(J)
```

```
5              CONTINUE
        Z(I)=SUM*X
4       CONTINUE
        IF(ABS((X1-X2)/X2).GT.D) GO TO  6
        GO TO 66
 6      DO 7 I=1,N
        Y(I)=Z(I)
7       CONTINUE
        X1=X2
          GO TO  2
66      DO 8 I=1,N
        IF(ABS(Z(I)).LE.0.000000001) GOTO 8
        IF(ABS((Y(I)-Z(I))/Z(I)).LE.D) GOTO 8
14              DO 10 J=1,N
                Y(J)=Z(J)
10              CONTINUE
                X1=X2
                GO TO  2
8       CONTINUE
        X = 0.0
        DO 11 I=1,N
        IF (ABS(X).LT.ABS(Y(I))) X=Y(I)
11      CONTINUE
        DO 12 I=1,N
        Y(I)=Y(I)/X
12      CONTINUE
        RETURN
        END
        SUBROUTINE EIGEN(A,N,AM,VEC,M1,D,XX,NP1)
        DIMENSION A(1),
      1 VEC(1),AM(1),XX(1),
      2 X(200),IG(200)
        MN=N
        NN = N
        CALL VALEIG(A,X,XM,NN,D,N)
        M=1
        DO 2 I=1,NN
        I3=(M-1)*N+I
        VEC(I3)=X(I)
        XX(I3)=VEC(I3)
2       CONTINUE
        AM(M) = XM
        DO 20 M=2,M1
        DO 4 I=1,NN
        IF(ABS(XX((M-2)*N+I)-1.0).LT.0.00001)IR=I
4       CONTINUE
```

```
      IG(M-1)=IR
      DO 5 I=1,NN
      XX((MN-M+2)*N+MN-I+1)=A((I-1)*N+IR)
5     CONTINUE
      DO 6 I=1,NN
      DO 6 J=1,NN
      I3=(J-1)*N+I
      I4=(M-2)*N+I
      I5=(MN-M+2)*N+MN-J+1
      A(I3)=A(I3)-XX(I4)*XX(I5)
6     CONTINUE
      DO 14 I=1,NN
      IF(I.EQ.IR) GO TO 14
      IF(I.GT.IR) K1=I-1
      IF(I.LE.IR) K1=I
           DO  7 J=1,NN
           IF(J.EQ.IR) GO TO 7
           IF(J.GT.IR) K2=J-1
      IF(J.LE.IR) K2=J
      A((K2-1)*N+K1)=A((J-1)*N+I)
7              CONTINUE
14    CONTINUE
      NN = NN-1
      M3=NN
      IF (M.NE.MN)  GO TO  11
      XM=A(1)
      X(1)=1.0
      GO TO  12
11    CALL VALEIG(A,X,XM,NN,D,N)
12    DO 13 I=1,NN
      XX((M-1)*N+I)=X(I)
13    CONTINUE
      AM(M)=XM
      M11 = M-1
      MM1=1000-M11
      DO 20 M8=MM1,999
      M12 = M3+1
      M2=1000-M8
      M22 =IG(M2)+1
      MM22=1000-M22
      MM12=1000-M12
      DO 15 I3=MM12,MM22
      I=1000-I3
      X(I)=X(I-1)
15    CONTINUE
      J=IG(M2)
```

```
         X(J)=0.0
         SUM=0.0
         DO 16 I= 1,M12
        SUM=SUM+XX((MN-M2+1)*N+MN-I+1)*X(I)
  16     CONTINUE
         XK=(AM(M2)-XM)/SUM
         DO 17 I= 1,M12
        X(I)=XX((M2-1)*N+I)-XK*X(I)
  17     CONTINUE
         SUM=0.0
         DO 18 I=1,M12
         IF (ABS(SUM).LT.ABS(X(I)))SUM=X(I)
  18     CONTINUE
         DO 19 I=1,M12
         X(I)=X(I)/SUM
  19     CONTINUE
         M3=M3+1
        IF(M2.NE.1)GO TO 20
         DO 3 I=1,M3
        VEC((M-1)*N+I)=X(I)
  3      CONTINUE
  20     CONTINUE
          RETURN
         END
C
         SUBROUTINE PRODAB(A,B,N)
         DIMENSION A(N,N),B(N,N),VEC(100,1)
C        (A)=(A)*(B)
         DO 1 I=1,N
         DO 2 J=1,N
         VEC(J,1)=0.0
         DO 2 K=1,N
         VEC(J,1)=VEC(J,1)+A(I,K)*B(K,J)
  2      CONTINUE
         DO 1 J=1,N
         A(I,J)=VEC(J,1)
  1      CONTINUE
         RETURN
         END
C
         SUBROUTINE  EIGPRIN2(VEC,XLAM,NJOINTS,
        1 NDOF,N,NV)
C
C        PRINTS RESULTS OF EIGENVALUE ANALYSIS:
C        FREQUENCY AND MODE SHAPE,INCORPORATING SUPPRESSED DOF'S.
C
```

```
        DIMENSION  VEC(N,NV),XLAM(NV)
        WRITE(6,30)
 30     FORMAT(///12X, 20H***EIGEN SOLUTION***)
        PI=3.1415927
        DO 150 IEIG=1,NV
           PAR=1.0/(2.0*PI)*SQRT(1.0/XLAM(IEIG))
        WRITE(6,60) IEIG, PAR
 60     FORMAT(///7X, I3, 1X, 12H FREQUENCY =,E14.6/
       1 8X, 29H------------------------------)
           WRITE(6,80)
 80        FORMAT(//10X,11H MODE SHAPE/
       1 10X,6H  NODE,5X, 37H RELATIVE DISPLACEMENTS(GLOBAL COORD))
           DO 90 IPOIN=1,NJOINTS
        IP=IPOIN*NDOF-NDOF
           WRITE(6,120) IPOIN, (VEC(IP+IDOF,IEIG),IDOF=1,NDOF)
 90        CONTINUE
 120    FORMAT(9X, I5, 3X, E10.3, 4X, E10.3, 4X, E10.3)
 150    CONTINUE
        RETURN
        END
C
        SUBROUTINE INVMX(A,N,D,L,M,B)
        DIMENSION A(1),B(1),L(1),M(1)
        I5=0
 3      I1=N-1
        DO 1 I=1,I1
        I2=I+1
        DO 2 J=I2,N
        I3=(I-1)*N+J
        I4=(J-1)*N+I
        B(J)=A(I3)
        A(I3)=A(I4)
 2      A(I4)=B(J)
 1      CONTINUE
        IF (I5.EQ.1) RETURN
        NK=-N
        DO 80 K=1,N
        NK=NK+N
        L(K)=K
        M(K)=K
        KK=NK+K
        BIGA=A(KK)
        DO 20 J=K,N
        IZ=N*(J-1)
        DO 20 I=K,N
        IJ=IZ+I
```

```
10    IF (ABS(BIGA)-ABS(A(IJ))) 15,20,20
15    BIGA=A(IJ)
      L(K)=I
      M(K)=J
20    CONTINUE
      J=L(K)
      IF (J-K) 35,35,25
25    KI=K-N
      DO 30 I=1,N
      KI=KI+N
      HOLD=-A(KI)
      JI=KI-K+J
      A(KI)=A(JI)
30    A(JI)=HOLD
35    I=M(K)
      IF (I-K) 45,45,38
38    JP=N*(I-1)
      DO 40 J=1,N
      JK=NK+J
      JI=JP+J
      HOLD=-A(JK)
      A(JK)=A(JI)
40    A(JI)=HOLD
45    IF (BIGA) 48,46,48
46    D=0.0
      RETURN
48    DO 55 I=1,N
      IF (I-K) 50,55,50
50    IK=NK+I
      A(IK)=A(IK)/(-BIGA)
55    CONTINUE
      DO 65 I=1,N
      IK=NK+I
      HOLD=A(IK)
      IJ=I-N
      DO 65 J=1,N
      IJ=IJ+N
      IF(I-K) 60,65,60
60    IF (J-K) 62,65,62
62    KJ=IJ-I+K
      A(IJ)=HOLD*A(KJ)+A(IJ)
65    CONTINUE
      KJ=K-N
      DO 75 J=1,N
      KJ=KJ+N
      IF (J-K) 70,75,70
```

```
70    A(KJ)=A(KJ)/BIGA
75    CONTINUE
      D=D*BIGA
      A(KK)=1.0/BIGA
80    CONTINUE
      K=N
100   K=(K-1)
      IF (K) 150,150,105
105   I=L(K)
      IF (I-K) 120,120,108
108   JQ=N*(K-1)
      JR=N*(I-1)
      DO 110 J=1,N
      JK=JQ+J
      HOLD=A(JK)
      JI=JR+J
      A(JK)=-A(JI)
110   A(JI)=HOLD
120   J=M(K)
      IF (J-K) 100,100,125
125   KI=K-N
      DO 130 I=1,N
      KI=KI+N
      HOLD=A(KI)
      JI=KI-K+J
      A(KI)=-A(JI)
130   A(JI)=HOLD
      GO TO 100
150   I5=1
      GO TO 3
      END
C
      SUBROUTINE SCPROD(A,M,N,CN)
      DIMENSION A(M,N)
C     (A)=CN*(A)
      DO 1 I=1,M
      DO 1 J=1,N
      A(I,J)=CN*A(I,J)
1     CONTINUE
      RETURN
      END
C
      SUBROUTINE NULL(A,M,N)
      DIMENSION A(M,N)
C     (A)=(NULL)
      DO 1 I=1,M
```

```
      DO 1 J=1,N
      A(I,J)=0.0
1     CONTINUE
      RETURN
      END
```

A 2.2 A Typical Input File for MAIN7DAT.DAT (Example 5.3)

```
4   4   3
1   1   1
3   1   0
4   1   0
5   1   0
2.0E11  7860.
1   2   .42E-5   .24E-2   2.
2   3   .42E-5   .24E-2   2.
3   4   .42E-5   .24E-2   3.
4   5   .42E-5   .24E-2   2.
1
2   1.5   .109E-2
```

A 2.3 A Typical Output File for OUT7.OUT (Example 5.3)

```
***BASIC PARAMETERS***

      NUMBER OF ELEMENTS . . . . 4
      NUMBER OF NODES WITH ZERO DISPLACEMENTS. . . . 4
      NUMBER OF EIGEN VALUES . . 3



DETAILS OF NODES WITH ZERO DISPLACEMENTS
      1     1     1
      3     1     0
      4     1     0
      5     1     0



      YOUNGS MODULUS=    0.200000E+12
      DENSITY=           0.786000E+04
```

ELEMENT DATA

I	J	SMA	CSA	LENGTH
1	2	0.42000001E-05	0.24000000E-02	0.20000000E+01
2	3	0.42000001E-05	0.24000000E-02	0.20000000E+01
3	4	0.42000001E-05	0.24000000E-02	0.30000001E+01
4	5	0.42000001E-05	0.24000000E-02	0.20000000E+01

NUMBER OF CONCENTRATED MASSES= 1

NODE	CON MASS	CON M.M.I.
2	0.1500E+01	0.1090E-02

EIGEN SOLUTION

1   FREQUENCY =   0.362960E+02

MODE SHAPE

NODE	RELATIVE	DISPLACEMENTS(GLOBAL COORD)
1	0.982E-12	-0.716E-12
2	0.100E+01	-0.199E+00
3	0.539E-12	0.713E+00
4	-0.394E-12	-0.271E+00
5	0.170E-12	0.151E+00

2   FREQUENCY =   0.704136E+02

MODE SHAPE

NODE	RELATIVE	DISPLACEMENTS(GLOBAL COORD)
1	0.880E-12	-0.474E-12
2	0.323E+00	0.230E+00
3	0.179E-11	-0.612E+00
4	0.827E-12	0.100E+01
5	-0.107E-11	-0.751E+00

3   FREQUENCY =   0.124660E+03

MODE SHAPE
  NODE   RELATIVE DISPLACEMENTS(GLOBAL COORD)
    1   0.191E-11   -0.826E-12
    2   0.128E+00   0.695E+00
    3   -0.357E-12   -0.739E+00
    4   0.215E-11   -0.531E+00
    5   0.206E-11   0.100E+01

Appendix 3. Computer Program for the Vibration of Rigid-jointed Plane Frames

A 3.1 The Computer Program

```
C      ********************************************
C      *VIBRATIONS OF RIGID JOINTED PLANE FRAMES*
C      ********************************************
C
       DIMENSION ST(60,60),SMAS(60,60),
      1 EV(20),VEC(60,20),XX(60,61)
       OPEN(5,FILE='MAIN8DAT.DAT',STATUS='OLD')
       OPEN(6,FILE='OUT8.OUT',STATUS='OLD')
       READ(5,*) NN,MEMS,NFIXD,NV
       NJ3=NN*3
       NP1=NJ3+1
       CALL XMAIN(ST,SMAS,NN,NJ3,NFIXD,NV,EV,MEMS,
      1 VEC,NP1,XX)
       CLOSE(5)
       CLOSE(6)
       STOP
       END
C
C
       SUBROUTINE XMAIN(ST,SMAS,NN,NJ3,NFIXD,NV,EV,MEMS,
      1 VEC,NP1,XX)
       DIMENSION ST(NJ3,NJ3),SMAS(NJ3,NJ3),
      1 EV(NV),VEC(NJ3,NV),XX(NJ3,NP1),
      2 NSUP(50),NPOSS(50),NXSUP(50),NYSUP(50),
      9 NTSUP(50),X(50),Y(50),
      3 DC(6,6),DCT(6,6),L(60),M(60),B(60),
      4 SK(6,6),SM(6,6),SMM(6,6)
C
       NJOINTS=NN
       NNODES=2
       NDF=3
C
       CALL NULL(ST,NJ3,NJ3)
       CALL NULL(SMAS,NJ3,NJ3)
       WRITE(6,101) NN,MEMS,NFIXD,NV
       WRITE(6,118)
       DO 1 II=1,NN
       READ(5,*) X(II), Y(II)
       WRITE(6,104) II, X(II), Y(II)
```

```
1      CONTINUE
       WRITE(6,105)
       DO 2 II=1,NFIXD
       READ(5,*) NPOSS(II),NXSUP(II),NYSUP(II),NTSUP(II)
       WRITE(6,106) NPOSS(II),NXSUP(II),NYSUP(II),NTSUP(II)
2      CONTINUE
       DO 777 II=1,NFIXD*3
       NSUP(II)=0
777    CONTINUE
       DO 77 II=1,NFIXD
       IF(NXSUP(II).NE.0)NSUP(II*3-2)=NPOSS(II)*3-2
       IF(NYSUP(II).NE.0)NSUP(II*3-1)=NPOSS(II)*3-1
       IF(NTSUP(II).NE.0)NSUP(II*3)=NPOSS(II)*3
77     CONTINUE
       READ(5,*) E,RHO
       WRITE(6,114) E,RHO
       WRITE(6,116)
C
       DO 5 MEM=1,MEMS
       READ(5,*) I, J, SMA, CSA
       WRITE(6,117) I, J, SMA, CSA
       XL=SQRT((X(J)-X(I))**2+(Y(J)-Y(I))**2)
       C=(X(J)-X(I))/XL
       S=(Y(J)-Y(I))/XL
       CALL NULL(DC,6,6)
       DC(1,1) = C
       DC(2,2) = C
       DC(4,4) = C
       DC(5,5) = C
       DC(1,2) = S
       DC(4,5) = S
       DC(2,1) =-S
       DC(5,4) =-S
       DC(3,3) =1.0
       DC(6,6) =1.0
       CALL MXTRAN(DCT,DC,6,6)
       C1=12.0*E*SMA*S**2/(XL**3)+C**2*CSA*E/XL
       C2=12.0*E*SMA*C*S/(XL**3)-C*S*CSA*E/XL
       C3=12.0*E*SMA*C**2/(XL**3)+S**2*CSA*E/XL
       C4= 6.0*E*SMA*S/(XL**2)
       C5= 6.0*E*SMA*C/(XL**2)
       C6= 4.0*E*SMA/XL
       SK(1,1)=C1
       SK(2,1)=-C2
       SK(3,1)=C4
       SK(4,1)=-C1
```

```
          SK(5,1)=C2
          SK(6,1)=C4
          SK(2,2)=C3
          SK(3,2)=-C5
          SK(4,2)=C2
          SK(5,2)=-C3
          SK(6,2)=-C5
          SK(3,3)=C6
          SK(4,3)=-C4
          SK(5,3)=C5
          SK(6,3)=0.5*C6
          SK(4,4)=C1
          SK(5,4)=-C2
          SK(6,4)=-C4
          SK(5,5)=C3
          SK(6,5)=C5
          SK(6,6)=C6
C
C
          CALL NULL(SM,6,6)
          C1=RHO*CSA*XL/6.0
          C2=RHO*CSA*XL/420.0
          SM(1,1)=2.0*C1
          SM(4,1)=C1
          SM(2,2)=156.0*C2
          SM(3,2)=-22.0*XL*C2
          SM(5,2)=54.0*C2
          SM(6,2)=13.0*XL*C2
          SM(2,3)=-22.0*XL*C2
          SM(3,3)=4.0*XL**2*C2
          SM(5,3)=-13.0*XL*C2
          SM(6,3)=-3.0*XL**2*C2
          SM(1,4)=C1
          SM(4,4)=2.0*C1
          SM(2,5)=54.0*C2
          SM(3,5)=-13.0*XL*C2
          SM(5,5)=156.0*C2
          SM(6,5)=22.0*XL*C2
          SM(2,6)=13.0*XL*C2
          SM(3,6)=-3.0*XL**2*C2
          SM(5,6)=22.0*XL*C2
          SM(6,6)=4.0*XL**2*C2
          CALL MXPROD(SMM,SM,DC,6,6,6)
          CALL MXPROD(SM,DCT,SMM,6,6,6)
          DO 99 III=1,6
          DO 99 JJJ=III+1,6
```

```
      SK(III,JJJ)=SK(JJJ,III)
  99  CONTINUE
      CALL ASSEMB(I,J,6,SK,ST,NJ3)
      CALL ASSEMB(I,J,6,SM,SMAS,NJ3)

  5   CONTINUE
C
      READ(5,*)NCONC
      WRITE(6,909)NCONC
 909  FORMAT(//45H NO. OF NODES WITH CONCENTRATED MASSES,
ETC.=,I2)
C
      IF(NCONC.EQ.0)GOTO 919
      DO 929 II=1,NCONC
      READ(5,*)NPOS,XMAS,XMMI
      WRITE(6,939)NPOS,XMAS,XMMI
 939  FORMAT(I5,4X,E12.4,4X,E12.4)
      I1=3*NPOS-2
      I7=I1+1
      I9=I7+1
      SMAS(I1,I1)=SMAS(I1,I1)+XMAS
      SMAS(I7,I7)=SMAS(I7,I7)+XMAS
      SMAS(I9,I9)=SMAS(I9,I9)+XMMI
 929  CONTINUE
 919  CONTINUE
      NFIXD2=NFIXD*NDF
      CALL SUPPRESS(ST,NSUP,NJ3,NFIXD2)
      CALL INVMX(ST,NJ3,0.0,L,M,B)
      CALL PRODAB(ST,SMAS,NJ3)
      D=0.001
      CALL EIGEN(ST,NJ3,EV,VEC,NV,D,XX,NP1)
      CALL EIGPRIN2(VEC,EV,NJOINTS,NDF,NJ3,NV)
C
C
C
 101  FORMAT(23H ***BASIC PARAMETERS***//
     1 5X, 29H NUMBER OF NODAL POINTS . . .,I3/
     2 5X, 29H NUMBER OF MEMBERS  . . . . .,I3/
     3 5X, 48H NUMBER OF NODES WITH ZERO DISPLACEMENTS . . .
.,I3/
     4 5X, 29H NUMBER OF EIGENVALUES. . . .,I3)
 102  FORMAT(///23H ***NODAL POINT DATA***//
     1 2X, 4HNODE, 6X, 1HX, 9X, 1HY)
 104  FORMAT(I5,2F10.3)
 105  FORMAT(////41H DETAILS OF NODES WITH ZERO DISPLACEMENTS)
 106  FORMAT(4I5)
```

```
    108   FORMAT(I5,F10.3)
    109   FORMAT(I5,3X,F10.3)
    111   FORMAT(2X,I2,3X,E13.6,4X,E13.6,4X,E13.6)
    112   FORMAT(2X,I2,5X,I2,4X,E13.6,4X,E13.6)
    114   FORMAT(//16H YOUNGS MODULUS=,E20.8/
         1 9H DENSITY=,7X,E20.8)
    116   FORMAT(//19H ***ELEMENT DATA***/
         1 4X, 1HI, 4X, 1HJ, 7X, 3HSMA, 11X, 3HCSA)
    117   FORMAT(2I5,4X,E10.3,4X,E10.3)
    118   FORMAT(//25H ***GLOBAL COORDINATES***/
         1 6H  NODE, 5X, 1HX, 9X, 1HY)
          RETURN
          END
C
          SUBROUTINE VALEIG(A,Y,X2,N,D,N1)
          DIMENSION A(1),
         2 Y(200),Z(200)
C         MODIFIED 20-11-78
          X1=100000.0
          DO 1 I=1,N
          Y(I)=1.0
    1     CONTINUE
    2     X2=-100000.0
          DO 3 I=1,N
          IF (X2.LT.Y(I))        X2=Y(I)
    3     CONTINUE
          X=1.0/X2
          DO 4 I=1,N
          SUM=0.0
                DO  5  J=1,N
          SUM=SUM+A((J-1)*N1+I)*Y(J)
    5           CONTINUE
          Z(I)=SUM*X
    4     CONTINUE
          IF(ABS((X1-X2)/X2).GT.D) GO TO   6
          GO TO 66
    6     DO 7 I=1,N
          Y(I)=Z(I)
    7     CONTINUE
          X1=X2
            GO TO   2
    66    DO 8 I=1,N
          IF(ABS(Z(I)).LE.0.000000001) GOTO 8
          IF(ABS((Y(I)-Z(I))/Z(I)).LE.D) GOTO 8
    14          DO 10 J=1,N
                  Y(J)=Z(J)
```

```
10                CONTINUE
                  X1=X2
                  GO TO  2
8      CONTINUE
       X = 0.0
       DO 11 I=1,N
      IF (ABS(X).LT.ABS(Y(I))) X=Y(I)
11     CONTINUE
       DO 12 I=1,N
       Y(I)=Y(I)/X
12     CONTINUE
       RETURN
       END
C
       SUBROUTINE EIGEN(A,N,AM,VEC,M1,D,XX,NP1)
       DIMENSION A(1),
      1 VEC(1),AM(1),XX(1),
      2 X(200),IG(200)
        MN=N
       NN = N
       CALL VALEIG(A,X,XM,NN,D,N)
        M=1
        DO 2 I=1,NN
       I3=(M-1)*N+I
       VEC(I3)=X(I)
       XX(I3)=VEC(I3)
2      CONTINUE
        AM(M) = XM
       DO 20 M=2,M1
       DO 4 I=1,NN
       IF(ABS(XX((M-2)*N+I)-1.0).LT.0.00001)IR=I
4      CONTINUE
        IG(M-1)=IR
       DO 5 I=1,NN
       XX((MN-M+2)*N+MN-I+1)=A((I-1)*N+IR)
5      CONTINUE
        DO 6 I=1,NN
        DO 6 J=1,NN
       I3=(J-1)*N+I
       I4=(M-2)*N+I
       I5=(MN-M+2)*N+MN-J+1
       A(I3)=A(I3)-XX(I4)*XX(I5)
6      CONTINUE
       DO 14 I=1,NN
       IF(I.EQ.IR) GO TO 14
        IF(I.GT.IR) K1=I-1
```

```
         IF(I.LE.IR) K1=I
              DO  7 J=1,NN
              IF(J.EQ.IR) GO TO 7
              IF(J.GT.IR) K2=J-1
         IF(J.LE.IR) K2=J
       A((K2-1)*N+K1)=A((J-1)*N+I)
7             CONTINUE
14     CONTINUE
       NN = NN-1
       M3=NN
       IF (M.NE.MN)  GO TO  11
       XM=A(1)
       X(1)=1.0
       GO TO  12
11     CALL VALEIG(A,X,XM,NN,D,N)
12     DO 13 I=1,NN
       XX((M-1)*N+I)=X(I)
13     CONTINUE
       AM(M)=XM
       M11 = M-1
       MM1=1000-M11
       DO 20 M8=MM1,999
        M12 = M3+1
       M2=1000-M8
        M22 =IG(M2)+1
       MM22=1000-M22
       MM12=1000-M12
       DO 15 I3=MM12,MM22
       I=1000-I3
        X(I)=X(I-1)
15     CONTINUE
       J=IG(M2)
       X(J)=0.0
       SUM=0.0
       DO 16 I= 1,M12
       SUM=SUM+XX((MN-M2+1)*N+MN-I+1)*X(I)
16     CONTINUE
       XK=(AM(M2)-XM)/SUM
       DO 17 I= 1,M12
       X(I)=XX((M2-1)*N+I)-XK*X(I)
17     CONTINUE
       SUM=0.0
       DO 18 I=1,M12
       IF (ABS(SUM).LT.ABS(X(I)))SUM=X(I)
18     CONTINUE
       DO 19 I=1,M12
```

```
         X(I)=X(I)/SUM
19       CONTINUE
         M3=M3+1
         IF(M2.NE.1)GO TO 20
          DO 3 I=1,M3
         VEC((M-1)*N+I)=X(I)
3        CONTINUE
20       CONTINUE
          RETURN
         END
C
         SUBROUTINE NULL(A,M,N)
         DIMENSION A(M,N)
         DO 1 I=1,M
         DO 1 J=1,N
         A(I,J)=0.0
  1      CONTINUE
         RETURN
         END
C
         SUBROUTINE PRODAB(A,B,N)
         DIMENSION A(N,N),B(N,N),VEC(100,1)
C        (A)=(A)*(B)
         DO 1 I=1,N
         DO 2 J=1,N
         VEC(J,1)=0.0
         DO 2 K=1,N
         VEC(J,1)=VEC(J,1)+A(I,K)*B(K,J)
  2      CONTINUE
         DO 1 J=1,N
         A(I,J)=VEC(J,1)
  1      CONTINUE
         RETURN
         END
C
         SUBROUTINE INVMX(A,N,D,L,M,B)
         DIMENSION A(1),B(1),L(1),M(1)
         I5=0
  3      I1=N-1
         DO 1 I=1,I1
         I2=I+1
         DO 2 J=I2,N
         I3=(I-1)*N+J
         I4=(J-1)*N+I
         B(J)=A(I3)
         A(I3)=A(I4)
```

```
 2      A(I4)=B(J)
 1      CONTINUE
        IF (I5.EQ.1) RETURN
        NK=-N
        DO 80 K=1,N
        NK=NK+N
        L(K)=K
        M(K)=K
        KK=NK+K
        BIGA=A(KK)
        DO 20 J=K,N
        IZ=N*(J-1)
        DO 20 I=K,N
        IJ=IZ+I
10      IF (ABS(BIGA)-ABS(A(IJ))) 15,20,20
15      BIGA=A(IJ)
        L(K)=I
        M(K)=J
20      CONTINUE
        J=L(K)
        IF (J-K) 35,35,25
25      KI=K-N
        DO 30 I=1,N
        KI=KI+N
        HOLD=-A(KI)
        JI=KI-K+J
        A(KI)=A(JI)
30      A(JI)=HOLD
35      I=M(K)
        IF (I-K) 45,45,38
38      JP=N*(I-1)
        DO 40 J=1,N
        JK=NK+J
        JI=JP+J
        HOLD=-A(JK)
        A(JK)=A(JI)
40      A(JI)=HOLD
45      IF (BIGA) 48,46,48
46      D=0.0
        RETURN
48      DO 55 I=1,N
        IF (I-K) 50,55,50
50      IK=NK+I
        A(IK)=A(IK)/(-BIGA)
55      CONTINUE
        DO 65 I=1,N
```

```
        IK=NK+I
        HOLD=A(IK)
        IJ=I-N
        DO 65 J=1,N
        IJ=IJ+N
        IF(I-K) 60,65,60
60      IF (J-K) 62,65,62
62      KJ=IJ-I+K
        A(IJ)=HOLD*A(KJ)+A(IJ)
65      CONTINUE
        KJ=K-N
        DO 75 J=1,N
        KJ=KJ+N
        IF (J-K) 70,75,70
70      A(KJ)=A(KJ)/BIGA
75      CONTINUE
        D=D*BIGA
        A(KK)=1.0/BIGA
80      CONTINUE
        K=N
100     K=(K-1)
        IF (K) 150,150,105
105     I=L(K)
        IF (I-K) 120,120,108
108     JQ=N*(K-1)
        JR=N*(I-1)
        DO 110 J=1,N
        JK=JQ+J
        HOLD=A(JK)
        JI=JR+J
        A(JK)=-A(JI)
110     A(JI)=HOLD
120     J=M(K)
        IF (J-K) 100,100,125
125     KI=K-N
        DO 130 I=1,N
        KI=KI+N
        HOLD=A(KI)
        JI=KI-K+J
        A(KI)=-A(JI)
130     A(JI)=HOLD
        GO TO 100
150     I5=1
        GO TO 3
        END
        SUBROUTINE MXPROD(AB,A,B,N,M,L)
```

```
      DIMENSION AB(N,L),A(N,M),B(M,L)
C     (AB)=(A)*(B)
      DO 1 I=1,N
      DO 1 J=1,L
      AB(I,J)=0.0
    1 CONTINUE
      DO 2 I=1,N
      DO 2 K=1,L
      DO 2 J=1,M
      AB(I,K)=AB(I,K)+A(I,J)*B(J,K)
    2 CONTINUE
      RETURN
      END
C
      SUBROUTINE MXTRAN(AT,A,M,N)
      DIMENSION AT(N,M),A(M,N)
C     (AT)=(A)T
      DO 1 I=1,M
      DO 1 J=1,N
      AT(J,I)=A(I,J)
    1 CONTINUE
      RETURN
      END
C
      SUBROUTINE ASSEMB(I,J,NDF2,ST,A,N)
      DIMENSION A(N,N),ST(NDF2,NDF2)
      NDF=NDF2/2
      I1=NDF*I-NDF
      J1=NDF*J-NDF
      DO 10 II=1,NDF
      DO 10 JJ=1,NDF
      MM=I1+II
      MN=I1+JJ
      NM=J1+II
      N9=J1+JJ
      A(MM,MN)=A(MM,MN)+ST(II,JJ)
      A(NM,MN)=A(NM,MN)+ST(II+NDF,JJ)
      A(MM,N9)=A(MM,N9)+ST(II,JJ+NDF)
      A(NM,N9)=A(NM,N9)+ST(II+NDF,JJ+NDF)
   10 CONTINUE
      RETURN
      END
C
      SUBROUTINE SUPPRESS(A,NSUP,N,NFIXD)
      DIMENSION A(N,N),NSUP(1)
      DO 10 II=1,NFIXD
```

```
         NS=NSUP(II)
         IF(NS.NE.0)A(NS,NS)=A(NS,NS)*1.0E12+1.0E12
  10     CONTINUE
         RETURN
         END
C
C
         SUBROUTINE   EIGPRIN2(VEC,XLAM,NJOINTS,NDOF,N,NV)
C
C        PRINTS RESULTS OF EIGENVALUE ANALYSIS:
C        FREQUENCY AND MODE SHAPE,INCORPORATING SUPPRESSED DOF'S.
C
         DIMENSION   VEC(N,NV),XLAM(NV)
         WRITE(6,30)
  30     FORMAT(///12X, 20H***EIGEN SOLUTION***)
         PI=3.1415927
         DO 150 IEIG=1,NV
           PAR=1.0/(2.0*PI)*SQRT(1.0/XLAM(IEIG))
         WRITE(6,60) IEIG, PAR
  60     FORMAT(///7X, I3, 1X, 12H FREQUENCY =,E14.6/
        1 8X, 29H---------------------------)
           WRITE(6,80)
  80       FORMAT(//10X,11H MODE SHAPE/
        1 10X,6H  NODE,5X, 37H RELATIVE DISPLACEMENTS(GLOBAL COORD))
           DO 90 IPOIN=1,NJOINTS
         IP=IPOIN*NDOF-NDOF
           WRITE(6,120) IPOIN, (VEC(IP+IDOF,IEIG),IDOF=1,NDOF)
  90       CONTINUE
 120     FORMAT(9X, I5, 3X, E10.3, 4X, E10.3, 4X, E10.3)
 150     CONTINUE
         RETURN
         END
C
```

A 3.2 A Typical Input File for MAIN8DAT.DAT (Example 6.3)

```
4  3  2  2
0.  0.
1.  4.
3.  4.
3.5 2.
1  1  1  1
4  1  1  1
2.0E11  7860
```

```
1   2   .27E-5   .14E-2
2   3   .18E-5   .11E-2
3   4   .18E-5   .11E-2
2
2   5.   .144E-1
3   8   .52E-1
```

A 3.3 A Typical Output File for OUT8.OUT (Example 6.3)

```
***BASIC PARAMETERS***

      NUMBER OF NODAL POINTS . . .   4
      NUMBER OF MEMBERS   . . . . .   3
      NUMBER OF NODES WITH ZERO DISPLACEMENTS . . . .   2
      NUMBER OF EIGENVALUES. . . .   2


***GLOBAL COORDINATES***
   NODE      X          Y
    1       0.000      0.000
    2       1.000      4.000
    3       3.000      4.000
    4       3.500      2.000




DETAILS OF NODES WITH ZERO DISPLACEMENTS
    1    1    1    1
    4    1    1    1



YOUNGS MODULUS=       0.20000000E+12
DENSITY=              0.78600001E+04



***ELEMENT DATA***
   I    J      SMA          CSA
   1    2    0.270E-05    0.140E-02
   2    3    0.180E-05    0.110E-02
   3    4    0.180E-05    0.110E-02



NO. OF NODES WITH CONCENTRATED MASSES, ETC.= 2
```

```
2      0.5000E+01      0.1440E-01
3      0.8000E+01      0.5200E-01
       ***EIGEN SOLUTION***



  1  FREQUENCY =  0.171567E+02
-----------------------------



     MODE SHAPE
       NODE      RELATIVE DISPLACEMENTS(GLOBAL COORD)
        1    0.673E-13      0.266E-14      0.652E-12
        2    0.100E+01     -0.246E+00     -0.234E+00
        3    0.998E+00      0.247E+00      0.229E+00
        4    0.708E-13     -0.197E-14      0.651E-12



  2  FREQUENCY =  0.669099E+02
-----------------------------



     MODE SHAPE
       NODE      RELATIVE DISPLACEMENTS(GLOBAL COORD)
        1   -0.226E-12      0.611E-14     -0.190E-11
        2    0.194E+00     -0.461E-01      0.100E+01
        3    0.198E+00      0.437E-01     -0.392E+00
        4    0.880E-13     -0.476E-14      0.455E-12
```

Appendix 4. Computer Program for the Vibration of Pin-jointed Space Trusses

A 4.1 The Computer Program

```
C     VIBRATION OF SPACE TRUSSES
      DIMENSION STIFF(60,60),XMS(60,60),
     8 VEC(60,20)
      CALL NULL(STIFF,60,60)
      CALL NULL(XMS,60,60)
      OPEN(5,FILE='MAIN9DAT.DAT',STATUS='OLD')
      OPEN(6,FILE='OUT9.OUT',STATUS='OLD')
      READ(5,*)NJ,NFIXD,MEMS,NP
      NJ3=3*NJ
      NP1=NJ3+1
      CALL MAIN9(STIFF,XMS,NJ,NFIXD,MEMS,NP,NJ3,NP1,VEC)
      CLOSE(5)
      CLOSE(6)
      STOP
      END
C
      SUBROUTINE MAIN9(STIFF,XMS,NJ,NFIXD,MEMS,NP,NJ3,NP1,VEC)
      DIMENSION STIFF(NJ3,NJ3),XMS(NJ3,NP1),
     8 SK(3,3),SM(6,6),
     7 VEC(NJ3,NP),XLAM(20),
     5 X(50),Y(50),Z(50),
     6 LVEC(60),MVEC(60),BVEC(60),
     4 NSUPR(50),NPOSS(50),NXSUP(50),NYSUP(50),NZSUP(50)
C
      NJOINTS=NJ
      NDF=3
      WRITE(6,101) NJ,NFIXD,MEMS,NP
 101  FORMAT(//23H ***BASIC PARAMETERS***//
     1 5X, 29H NUMBER OF NODAL POINTS * * *,I5/
     2 5X, 50H NUMBER OF NODES WITH ZERO DISPLACEMENTS * * * *
     *,I5/
     3 5X, 29H NUMBER OF MEMBERS* * * * *,I5/
     4 5X, 29H NUMBER OF EIGEN VALUES REQD*,I5)
      PI=3.141592654
      WRITE(6,200)
 200  FORMAT(///12X, 1HX, 19X, 1HY, 19X, 1HZ)
      DO 1 I=1,NJ
      READ(5,*)X(I),Y(I),Z(I)
      WRITE(6,102)X(I),Y(I),Z(I)
```

```
1     CONTINUE
      WRITE(6,201)
201   FORMAT(///4X, 1HI, 4X, 1HJ, 10X, 4HAREA, 17X, 1HE, 18X,
3HRHO)
      DO 3 MEM=1,MEMS
      READ(5,*)I,J,A,E,RHO
      WRITE(6,103)I,J,A,E,RHO
102   FORMAT(3E20.8)
103   FORMAT(2I5,3E20.8)
      CALL NULL(SM,6,6)
      XL=SQRT((X(J)-X(I))**2+(Y(J)-Y(I))**2+(Z(J)-Z(I))**2)
      COSX=(X(J)-X(I))/XL
      COSY=(Y(J)-Y(I))/XL
      COSZ=(Z(J)-Z(I))/XL
      CN=A*E/XL
      RAL=RHO*A*XL
      SK(1,1)=COSX*COSX*CN
      SK(1,2)=COSX*COSY*CN
      SK(1,3)=COSX*COSZ*CN
      SK(2,1)=COSX*COSY*CN
      SK(2,2)=COSY*COSY*CN
      SK(2,3)=COSY*COSZ*CN
      SK(3,1)=COSX*COSZ*CN
      SK(3,2)=COSY*COSZ*CN
      SK(3,3)=COSZ*COSZ*CN
      DO 4 II=1,6
      SM(II,II)=RAL/3.0
4     CONTINUE
      DO 5 II=1,3
      SM(II+3,II)=RAL/6.0
      SM(II,II+3)=RAL/6.0
5     CONTINUE
      I1=3*I-3
      J1=3*J-3
      DO 6 II=1,3
      DO 6 JJ=1,3
      MM=I1+II
      MN=I1+JJ
      NM=J1+II
      NN=J1+JJ
      STIFF(MM,MN)=STIFF(MM,MN)+SK(II,JJ)
      STIFF(NM,NN)=STIFF(NM,NN)+SK(II,JJ)
      STIFF(MM,NN)=STIFF(MM,NN)-SK(II,JJ)
      STIFF(NM,MN)=STIFF(NM,MN)-SK(II,JJ)
      XMS(MM,MN)=XMS(MM,MN)+SM(II,JJ)
      XMS(MM,NN)=XMS(MM,NN)+SM(II,JJ+3)
```

```
         XMS(NM,MN)=XMS(NM,MN)+SM(II+3,JJ)
         XMS(NM,NN)=XMS(NM,NN)+SM(II+3,JJ+3)
  6      CONTINUE
  3      CONTINUE
         WRITE(6,202)
 202     FORMAT(///41H DETAILS OF NODES WITH ZERO DISPLACEMENTS)
         DO 7 II=1,NFIXD
         READ(5,*)NPOSS(II),NXSUP(II),NYSUP(II),NZSUP(II)
         WRITE(6,104)NPOSS(II),NXSUP(II),NYSUP(II),NZSUP(II)
  7      CONTINUE
 104     FORMAT(4I10)
         DO 777 II=1,NFIXD*3
         NSUPR(II)=0
 777     CONTINUE
         DO 77 II=1,NFIXD
         IF(NXSUP(II).NE.0)NSUPR(II*3-2)=NPOSS(II)*3-2
         IF(NYSUP(II).NE.0)NSUPR(II*3-1)=NPOSS(II)*3-1
         IF(NZSUP(II).NE.0)NSUPR(II*3)=NPOSS(II)*3
  77     CONTINUE
         READ(5,*) NCONC
         WRITE(6,106) NCONC
 106     FORMAT(//31H NUMBER OF CONCENTRATED MASSES=,I2)
         IF(NCONC.EQ.0)  GOTO 107
         WRITE(6,111)
 111     FORMAT(//2X, 4HNODE, 3X,    8HCON.MASS)
         DO 108 II=1,NCONC
         READ(5,*) NPOS, XMCONC
         WRITE(6,109) NPOS, XMCONC
 109     FORMAT(I5,1X,E12.4)
         I1=3*NPOS-2
         J1=3*NPOS-1
         K1=3*NPOS
         XMS(I1,I1)=XMS(I1,I1)+XMCONC
         XMS(J1,J1)=XMS(J1,J1)+XMCONC
         XMS(K1,K1)=XMS(K1,K1)+XMCONC
 108     CONTINUE
 107     CONTINUE
         NFIXD2=NFIXD*NDF
         CALL SUPPRESS(STIFF,NSUPR,NJ3,NFIXD2)
         CALL INVMX(STIFF,NJ3,0.0,LVEC,MVEC,BVEC)
         CALL PRODAB(STIFF,XMS,NJ3)
         CALL NULL(XMS,NJ3,NJ3)
         D=0.001
         CALL EIGEN(STIFF,NJ3,XLAM,VEC,NP,D,XX,NP1)
         CALL EIGPRIN2(VEC,XLAM,NJOINTS,NDF,NJ3,NP)
 105     FORMAT(E20.8)
```

```
      RETURN
      END
C
      SUBROUTINE NULL(A,M,N)
      DIMENSION A(M,N)
C     (A)=(NULL)
      DO 1 I=1,M
      DO 1 J=1,N
      A(I,J)=0.0
  1   CONTINUE
      RETURN
      END
C
      SUBROUTINE INVMX(A,N,D,L,M,B)
      DIMENSION A(1),B(1),L(1),M(1)
      I5=0
  3   I1=N-1
      DO 1 I=1,I1
      I2=I+1
      DO 2 J=I2,N
      I3=(I-1)*N+J
      I4=(J-1)*N+I
      B(J)=A(I3)
      A(I3)=A(I4)
  2   A(I4)=B(J)
  1   CONTINUE
      IF (I5.EQ.1) RETURN
      NK=-N
      DO 80 K=1,N
      NK=NK+N
      L(K)=K
      M(K)=K
      KK=NK+K
      BIGA=A(KK)
      DO 20 J=K,N
      IZ=N*(J-1)
      DO 20 I=K,N
      IJ=IZ+I
  10  IF (ABS(BIGA)-ABS(A(IJ))) 15,20,20
  15  BIGA=A(IJ)
      L(K)=I
      M(K)=J
  20  CONTINUE
      J=L(K)
      IF (J-K) 35,35,25
  25  KI=K-N
```

```
        DO 30 I=1,N
        KI=KI+N
        HOLD=-A(KI)
        JI=KI-K+J
        A(KI)=A(JI)
30      A(JI)=HOLD
35      I=M(K)
        IF (I-K) 45,45,38
38      JP=N*(I-1)
        DO 40 J=1,N
        JK=NK+J
        JI=JP+J
        HOLD=-A(JK)
        A(JK)=A(JI)
40      A(JI)=HOLD
45      IF (BIGA) 48,46,48
46      D=0.0
        RETURN
48      DO 55 I=1,N
        IF (I-K) 50,55,50
50      IK=NK+I
        A(IK)=A(IK)/(-BIGA)
55      CONTINUE
        DO 65 I=1,N
        IK=NK+I
        HOLD=A(IK)
        IJ=I-N
        DO 65 J=1,N
        IJ=IJ+N
        IF(I-K) 60,65,60
60      IF (J-K) 62,65,62
62      KJ=IJ-I+K
        A(IJ)=HOLD*A(KJ)+A(IJ)
65      CONTINUE
        KJ=K-N
        DO 75 J=1,N
        KJ=KJ+N
        IF (J-K) 70,75,70
70      A(KJ)=A(KJ)/BIGA
75      CONTINUE
        D=D*BIGA
        A(KK)=1.0/BIGA
80      CONTINUE
        K=N
100     K=(K-1)
        IF (K) 150,150,105
```

```
105    I=L(K)
       IF (I-K) 120,120,108
108    JQ=N*(K-1)
       JR=N*(I-1)
       DO 110 J=1,N
       JK=JQ+J
       HOLD=A(JK)
       JI=JR+J
       A(JK)=-A(JI)
110    A(JI)=HOLD
120    J=M(K)
       IF (J-K) 100,100,125
125    KI=K-N
       DO 130 I=1,N
       KI=KI+N
       HOLD=A(KI)
       JI=KI-K+J
       A(KI)=-A(JI)
130    A(JI)=HOLD
       GO TO 100
150    I5=1
       GO TO 3
       END
C
       SUBROUTINE VALEIG(A,Y,X2,N,D,N1)
       DIMENSION A(1),Y(1),
      2 Z(200)
       X1=100000.0
       DO 1 I=1,N
       Y(I)=1.0
1      CONTINUE
       X2=-100000.0
2      CONTINUE
       DO 4 I=1,N
       SUM=0.0
            DO 5  J=1,N
       SUM=SUM+A((J-1)*N1+I)*Y(J)
5              CONTINUE
       Z(I)=SUM
4      CONTINUE
       X2=0.0
       DO 21 I=1,N
       IF(ABS(X2).LT.ABS(Z(I))) X2=Z(I)
21     CONTINUE
       DO 22 I=1,N
       Y(I)=Z(I)/X2
```

```
  22    CONTINUE
        IF(ABS((X1-X2)/X2).GT.D) GO TO   6
        GO TO 66
   6    CONTINUE
        X1=X2
          GO TO  2
  66    CONTINUE
        X = 0.0
        DO 11 I=1,N
        IF (ABS(X).LT.ABS(Y(I))) X=Y(I)
  11    CONTINUE
        DO 12 I=1,N
        Y(I)=Y(I)/X
  12    CONTINUE
        RETURN
        END
C
        SUBROUTINE EIGEN(A,N,AM,VEC,M1,D,XX,NP1)
        DIMENSION A(1),
       1 VEC(1),AM(1),XX(1),
       2 X(200),IG(200)
          MN=N
        NN = N
        CALL VALEIG(A,X,XM,NN,D,N)
         M=1
         DO 2 I=1,NN
        I3=(M-1)*N+I
        VEC(I3)=X(I)
        XX(I3)=VEC(I3)
   2    CONTINUE
        AM(M) = XM
        IF(M1.LT.2) RETURN
        DO 20 M=2,M1
        DO 4 I=1,NN
        IF(ABS(XX((M-2)*N+I)-1.0).LT.0.00001)IR=I
   4    CONTINUE
        IG(M-1)=IR
        DO 5 I=1,NN
        XX((MN-M+2)*N+MN-I+1)=A((I-1)*N+IR)
   5    CONTINUE
        DO 6 I=1,NN
        DO 6 J=1,NN
        I3=(J-1)*N+I
        I4=(M-2)*N+I
        I5=(MN-M+2)*N+MN-J+1
        A(I3)=A(I3)-XX(I4)*XX(I5)
```

```
6       CONTINUE
        DO 9 I=1,NN
        IF(I.EQ.IR) GO TO 8
         IF(I.GT.IR) K1=I-1
         IF(I.LE.IR) K1=I
               DO  7 J=1,NN
                  IF(J.EQ.IR) GO TO 7
                  IF(J.GT.IR) K2=J-1
         IF(J.LE.IR) K2=J
        A((K2-1)*N+K1)=A((J-1)*N+I)
7                 CONTINUE
8       CONTINUE
9       CONTINUE
        NN = NN-1
         M3=NN
         IF (M.NE.MN)  GO TO  11
        XM=A(1)
         X(1)=1.0
         GO TO  12
11      CALL VALEIG(A,X,XM,NN,D,N)
12      DO 13 I=1,NN
        XX((M-1)*N+I)=X(I)
13      CONTINUE
         AM(M)=XM
         M11 = M-1
        MM1=1000-M11
        DO 20 M8=MM1,999
         M12 = M3+1
        M2=1000-M8
         M22 =IG(M2)+1
        MM22=1000-M22
        MM12=1000-M12
        IF(M12.LT.M22) GO TO 14
        DO 15 I3=MM12,MM22
        I=1000-I3
         X(I)=X(I-1)
15      CONTINUE
14      CONTINUE
         J=IG(M2)
         X(J)=0.0
         SUM=0.0
         DO 16 I= 1,M12
        SUM=SUM+XX((MN-M2+1)*N+MN-I+1)*X(I)
16      CONTINUE
         XK=(AM(M2)-XM)/SUM
         DO 17 I= 1,M12
```

```
            X(I)=XX((M2-1)*N+I)-XK*X(I)
   17       CONTINUE
            SUM=0.0
            DO 18 I=1,M12
            IF (ABS(SUM).LT.ABS(X(I)))SUM=X(I)
   18       CONTINUE
            DO 19 I=1,M12
            X(I)=X(I)/SUM
   19       CONTINUE
            M3=M3+1
            IF(M2.NE.1)GO TO 20
            DO 3 I=1,M3
            VEC((M-1)*N+I)=X(I)
    3       CONTINUE
   20       CONTINUE
            RETURN
            END
C
            SUBROUTINE SUPPRESS(A,NSUP,N,NFIXD)
            DIMENSION A(N,N),NSUP(1)
            DO 10 II=1,NFIXD
            NS=NSUP(II)
            IF(NS.NE.0)A(NS,NS)=A(NS,NS)*1.0E12+1.0E12
   10       CONTINUE
            RETURN
            END
C
C
            SUBROUTINE   EIGPRIN2(VEC,XLAM,NJOINTS,NDOF,N,NV)
C
C           PRINTS RESULTS OF EIGENVALUE ANALYSIS:
C           FREQUENCY AND MODE SHAPE,INCORPORATING SUPPRESSED DOF'S.
C
            DIMENSION   VEC(N,NV),XLAM(NV)
            WRITE(6,30)
   30       FORMAT(///12X, 20H***EIGEN SOLUTION***)
            PI=3.1415927
            DO 150 IEIG=1,NV
              PAR=1.0/(2.0*PI)*SQRT(1.0/XLAM(IEIG))
            WRITE(6,60) IEIG, PAR
   60       FORMAT(///7X, I3, 1X, 12H FREQUENCY =,E14.6/
          1 8X, 29H----------------------------)
              WRITE(6,80)
   80       FORMAT(//10X,11H MODE SHAPE/
          1 10X,6H  NODE,5X, 37H RELATIVE DISPLACEMENTS(GLOBAL COORD))
              DO 90 IPOIN=1,NJOINTS
```

```
        IP=IPOIN*NDOF-NDOF
          WRITE(6,120)  IPOIN,(VEC(IP+IDOF,IEIG),IDOF=1,NDOF)
  90      CONTINUE
 120    FORMAT(9X,  I5,  3X,  E10.3,  4X,  E10.3,  4X,  E10.3)
 150    CONTINUE
        RETURN
        END
C
        SUBROUTINE  PRODAB(A,B,N)
        DIMENSION  A(N,N),B(N,N),VEC(500,1)
C       (A)=(A)*(B)
        DO 1 I=1,N
        DO 2 J=1,N
        VEC(J,1)=0.0
        DO 2 K=1,N
        VEC(J,1)=VEC(J,1)+A(I,K)*B(K,J)
  2     CONTINUE
        DO 1 J=1,N
        A(I,J)=VEC(J,1)
  1     CONTINUE
        RETURN
        END
```

A 4.2 A Typical Input File for MAIN9DAT.DAT (Example 7.2)

```
4   3   3   1
0.   0.   0.
12.   0.   0.
6.   10.39  0.
6.   5.195   15.25
1   4   .12E-1   3.0E7   .735E-3
2   4   .12E-1   3.0E7   .735E-3
3   4   .12E-1   3.0E7   .735E-3
1   1   1   1
2   1   1   1
3   1   1   1
1
4   .518E-2
```

A 4.3 A Typical Output File for OUT9.OUT (Example 7.2)

BASIC PARAMETERS

NUMBER OF NODAL POINTS * * *   4
NUMBER OF NODES WITH ZERO DISPLACEMENTS * * * * *   3
NUMBER OF MEMBERS* * * * * *   3
NUMBER OF EIGEN VALUES REQD*   1

X	Y	Z
0.00000000E+00	0.00000000E+00	0.00000000E+00
0.12000001E+02	0.00000000E+00	0.00000000E+00
0.60000002E+01	0.10390000E+02	0.00000000E+00
0.60000002E+01	0.51950001E+01	0.15250000E+02

I	J	AREA	E	RHO
1	4	0.12000001E-01	0.30000001E+08	0.73500001E-03
2	4	0.12000001E-01	0.30000001E+08	0.73500001E-03
3	4	0.12000001E-01	0.30000001E+08	0.73500001E-03

DETAILS OF NODES WITH ZERO DISPLACEMENTS

1	1	1	1
2	1	1	1
3	1	1	1

NUMBER OF CONCENTRATED MASSES= 1

NODE	CON.MASS
4	0.5180E-02

```
***EIGEN SOLUTION***



1  FREQUENCY =   0.155727E+03
-------------------------------


   MODE SHAPE
     NODE       RELATIVE  DISPLACEMENTS (GLOBAL  COORD)
       1     0.141E-11      0.162E-11       0.551E-12
       2     0.609E-12     -0.684E-12      -0.236E-12
       3     0.227E-10      0.809E-12      -0.273E-12
       4     0.100E+01      0.632E+00      -0.580E-01
```

Appendix 5. Computer Program for the Vibration of Rigid-jointed Space Frames

A 5.1 The Computer Program

```
C       VIBRATIONS OF RIGID JOINTED SPACE FRAMES
        DIMENSION STIFF(48,48),XMS(48,48),
       9ST(24,24),XSM(24,24),
       8 VEC(24,6),XX(24,25)
C         MODIFIED 21sT JULY 1990 FOR I.B.M. PC
C         COPYRIGHT OF DR.C.T.F.ROSS
        open(5,file='vibdat3d.dat',status='old')
        open(6,file='out11.out',status='old')
        READ(5,*) NJ,NF,NFIXD,NIMK,N,NP,NMATL,MEMS
        NJ6=6*NJ
        NP1=N+1
        CALL VBRJSF(STIFF,XMS,ST,XSM,XX,VEC,
       9 NJ,NF,NFIXD,NIMK,N,NMATL,MEMS,NJ6,NP,NP1)
        close(5)
        close(6)
        STOP
C
        END
        SUBROUTINE VBRJSF(STIFF,XMS,ST,XSM,XX,VEC,
       9 NJ,NF,NFIXD,NIMK,N,NMATL,MEMS,NJ6,NP,NP1)
        DIMENSION DC(12,12),DCT(12,12),SK(12,12),SKX(12,12),
       4SM(12,12),SMX(12,12),QPT(24),
       3 X(24),Y(24),Z(24),
       3 XLAM2(24),VECTOR(24),XLAM(24),
       5 STIFF(NJ6,NJ6),XMS(NJ6,NJ6),ST(N,N),XSM(N,N),
       6 VEC(N,NP),XX(N,NP1),
       1 DENS(10),NPOSN(24),XMASS(24),
       6 LVEC(24,1),MVEC(24,1),BVEC(24,1),
       1 NSUPR(24),
       1 ELAST(10),RIGD(10),AREA(10),
       1 XIN(10),YIN(10),ZIN(10),XJP(10),NMA(50),IJK(200)
103     FORMAT(3E12.5)
104     FORMAT(3I10)
105     FORMAT(I10)
106     FORMAT(8E12.5)
107     FORMAT(6E18.11)
108     FORMAT(I10,F10.0)
109     FORMAT(I10,E15.4)
111     FORMAT(I3,4E15.4,2I3)
```

```
200   FORMAT(23H***BASIC PARAMETERS*** //
  1 5X, 47H NUMBER OF ACTUAL (NON-ZERO) NODAL POINTS . . .,I5/
  1 5X, 39H NUMBER OF FIXED (ZERO) NODES . .   .,I5/
  2 5X, 49H NO. OF ZERO DISPLTS. AT PIN-JOINTS, ETC. . . .
    .,I5/
  4 5X, 29H NUMBER OF IMAGINARY NODES. .,I5/
  5 5X, 29H NUMBER OF FREE DISPLACEMENTS,I5/
  5 5X, 29H NUMBER OF FREQUENCIES .  .  ,I5/
  6 5X, 29H NUMBER OF MEMBER TYPES .  .,I5/
  7 5X, 29H NUMBER OF MEMBERS  . . . . .,I5)
210   FORMAT(///8X, 12H COORDINATES//
  1 4X, 2HXI, 9X, 2HYI, 10X, 2HZI)
220   FORMAT(//4X, 3HRHO, 11X, 1HE, 11X, 1HG,
  1 10X, 4HAREA, 9X, 2HTC, 11X, 2HIY, 10X, 2HIZ,10X,2HJP)
230   FORMAT(30HNUMBER OF CONCENTRATED LOADS =,I3)
240   FORMAT(53H(FINISHED POSITIONS) AND VALUES OF CONCENTRATED
      LOADS)
250   FORMAT(53HNODAL POSITIONS & OTHER DETAILS OF ZERO
      DISPLACEMENTS)
303   FORMAT(F10.0)
304   FORMAT(E20.8)
      WRITE(6,200) NJ,NF,NFIXD,NIMK,N,NP,NMATL,MEMS
      NODELM=2
      NDF=6
      WRITE(6,210)
      NNIMK=NJ+NIMK
      DO 122 I=1,NNIMK
      READ(5,*)X(I),Y(I),Z(I)
      WRITE(6,103)X(I),Y(I),Z(I)
122   CONTINUE
      WRITE(6,220)
      DO 123 I=1,NMATL
      READ(5,*) DENS(I),ELAST(I),RIGD(I),AREA(I),XIN(I),
  9 YIN(I),ZIN(I),XJP(I)
      WRITE(6,106)DENS(I),ELAST(I),RIGD(I),AREA(I),XIN(I),
  9 YIN(I),ZIN(I),XJP(I)
123   CONTINUE
      WRITE(6,223)
      DO 99 MEM=1,MEMS
      READ(5,*)I,J,K
      WRITE(6,104)I,J,K
      M8=3*MEM
      IJK(M8-2)=I
      IJK(M8-1)=J
      IJK(M8)=K
      IF (NMATL.EQ.1)NMA(1)=1
```

```
         IF (NMATL.EQ.1)GOTO 99
         READ(5,*)NMA(MEM)
         WRITE(6,999)NMA(MEM)
   99    CONTINUE
  999    FORMAT(14HMATERIAL TYPE=,I5)
         IF(NFIXD.EQ.0)GOTO 707
         DO 77 II=1,NFIXD
         NSUPR(II)=0
   77    CONTINUE
         WRITE(6,666)
  666    FORMAT(49HDETAILS OF SUPPRESSED DISPLTS. AT PIN-JOINTS,ETC)
         DO 777 II=1,NFIXD
         READ(5,*)NSUPR(II)
         WRITE(6,717)NSUPR(II)
  717    FORMAT(I5)
  777    CONTINUE
  707    CONTINUE
         READ(5,*)NCONC
         WRITE(6,939)NCONC
  939    FORMAT(38HNO. OF NODES WITH CONCENTRATED MASSES=,I5)
         IF(NCONC.LE.0)GOTO 949
         DO 959 II=1,NCONC
C        TYPE IN THE FINISHED POSITION OF THE (U) DISPLACEMENT
C        OF THE MASS & THE VALUE OF THE MASS
         READ(5,*)NPOSN(II),XMASS(II)
         WRITE(6,969)NPOSN(II),XMASS(II)
  969    FORMAT(52HTHE FINISHED U DISPT. POSITION OF MASS ITS VALUE
        ARE, 9 I5,E20.8)
  959    CONTINUE
  949    CONTINUE
         CALL NULL(STIFF,NJ6,NJ6)
         CALL NULL(XMS,NJ6,NJ6)
         IDS=0
         JDS=0
         IMAX=0
  223    FORMAT(//9X, 1HI, 9X, 1HJ, 9X, 1HK)
         DO 131 MEM=1,MEMS
         WRITE(*,755)MEM
  755    FORMAT('  ELEMENT NO.',I5,' UNDER COMPUTATION')
         ISUP=0
         JSUP=0
         KSUP=0
         IF(NMATL.EQ.1) THEN
         MATL=1
         ELSE
         MATL=NMA(MEM)
```

```
      END IF
      E=ELAST(MATL)
      RHO=DENS(MATL)
      G=RIGD(MATL)
      CSA=AREA(MATL)
      TC=XIN(MATL)
      SMAY=YIN(MATL)
      SMAZ=ZIN(MATL)
      XPOL=XJP(MATL)
      M8=3*MEM
      I=IJK(M8-2)
      J=IJK(M8-1)
      K=IJK(M8)
      CALL NULL(DC,12,12)
      CALL NULL(DCT,12,12)
      CALL NULL(SK,12,12)
      CALL NULL(SKX,12,12)
      CALL NULL (SM,12,12)
      CALL NULL (SMX,12,12)
      IF (I.LT.0) CALL ISOLATE(I,ISUP,IX,IY,IZ,IAX,IAY,IAZ)
      IF (J.LT.0) CALL ISOLATE(J,JSUP,JX,JY,JZ,JAX,JAY,JAZ)
      IF (I.GT.999)CALL ELIM6(I,ISUP,IX,IY,IZ,IAX,IAY,IAZ)
      IF (J.GT.999)CALL ELIM6(J,JSUP,JX,JY,JZ,JAX,JAY,JAZ)
      IF (I.EQ.0) THEN
C     COORDS OF FIXED I NODE
      READ (5,*)XI,YI,ZI
      WRITE(6,103)XI,YI,ZI
      ELSE
      XI=X(I)
      YI=Y(I)
      ZI=Z(I)
      END IF
      IF (J.EQ.0) THEN
C     COORDS OF FIXED J NODE
      READ(5,*)XJ,YJ,ZJ
      WRITE(6,103)XJ,YJ,ZJ
      ELSE
      XJ=X(J)
      YJ=Y(J)
      ZJ=Z(J)
      END IF
      IF (K.EQ.0) THEN
C     COORDS OF FIXED K NODE
      READ(5,*)XK,YK,ZK
      WRITE(6,103)XK,YK,ZK
      ELSE
```

```
          XK=X(K)
          YK=Y(K)
          ZK=Z(K)
          END IF
          XL=(XJ-XI)**2+(YJ-YI)**2+(ZJ-ZI)**2
          XL=SQRT(XL)
          CALL DIRCOS(DC,XI,XJ,XK,YI,YJ,YK,ZI,ZJ,ZK)
          CALL STMS(SK,SM,DC,DCT,SKX,SMX,XL,
         1CSA,TC,SMAY,SMAZ,XPOL,RHO,E,G)
          CALL MXTRAN(DCT,DC,12,12)
          CALL MXPROD(SKX,SK,DC,12,12,12)
          CALL MXPROD(SK,DCT,SKX,12,12,12)
          CALL MXPROD(SKX,SM,DC,12,12,12)
          CALL MXPROD(SM,DCT,SKX,12,12,12)
          CALL ASSKM(STIFF,XMS,SK,SM,I,J,NJ6,
         1ISUP,JSUP,IX,IY,IZ,IAX,IAY,IAZ,
         2JX,JY,JZ,JAX,JAY,JAZ,IDS,IMAX)
  131     CONTINUE
          IF(NFIXD.EQ.0)GOTO 737
          DO 747 II=1,NFIXD
          N9=NSUPR(II)
          IF(N9.NE.0)GOTO 747
          STIFF(N9,N9)=STIFF(N9,N9)*1.0E12
  747     CONTINUE
  737     IF(NCONC.EQ.0)GOTO 767
          DO 787 II=1,NCONC
          N9=NPOSN(II)
          XMS(N9,N9)=XMS(N9,N9)+XMASS(II)
          XMS(N9+1,N9+1)=XMS(N9+1,N9+1)+XMASS(II)
          XMS(N9+2,N9+2)=XMS(N9+2,N9+2)+XMASS(II)
  787     CONTINUE
  767     CONTINUE
          DO 132 II=1,N
          DO 132 JJ=1,N
          ST(II,JJ)=STIFF(II,JJ)
          XSM(II,JJ)=XMS(II,JJ)
  132     CONTINUE
          DETL=0.0
          CALL INVMX(ST,N,DETL,LVEC,MVEC,BVEC)
          CALL PRODAB(ST,XSM,N)
          D=0.001
          WRITE(6,105) NP
          NP1=N+1
          CALL EIGEN(ST,N,XLAM,VEC,NP,D,XX,NP1)
          CALL EIGPT(ST,N,XLAM,VEC,NP)
  C
```

```
      RETURN
      END
C
      SUBROUTINE STMS(SK,SM,DC,DCT,SKX,SMX,XL,
     1 A,TC,SMAY,SMAZ,XPOL,RHO,E,G)
      DIMENSION SK(12,12),SM(12,12),DC(12,12),
     2DCT(12,12),SKX(12,12),SMX(12,12)
      SK(1,1)=A*E/XL
      SK(7,7)=SK(1,1)
      SK(7,1)=-SK(1,1)
      SK(1,7)=-SK(1,1)
      SK(2,2)=12.*E*SMAZ/(XL**3)
      SK(8,8)=SK(2,2)
      SK(8,2)=-SK(2,2)
      SK(2,8)=-SK(2,2)
      SK(3,3)=12.*E*SMAY/(XL**3)
      SK(9,9)=SK(3,3)
      SK(9,3)=-SK(3,3)
      SK(3,9)=-SK(3,3)
      SK(4,4)=G*TC/XL
      SK(10,10)=SK(4,4)
      SK(10,4)=-SK(4,4)
      SK(4,10)=-SK(4,4)
      SK(5,5)=4.*E*SMAY/XL
      SK(11,11)=SK(5,5)
      SK(11,5)=SK(5,5)/2.
      SK(5,11)=SK(11,5)
      SK(6,6)=4.*E*SMAZ/XL
      SK(12,12)=SK(6,6)
      SK(12,6)=SK(6,6)/2.
      SK(6,12)=SK(12,6)
      SK(6,2)=6.*E*SMAZ/(XL**2)
      SK(2,6)=SK(6,2)
      SK(12,2)=SK(6,2)
      SK(2,12)=SK(6,2)
      SK(8,6)=-SK(6,2)
      SK(6,8)=-SK(6,2)
      SK(12,8)=-SK(6,2)
      SK(8,12)=-SK(6,2)
      SK(9,5)=6.*E*SMAY/(XL**2)
      SK(5,9)=SK(9,5)
      SK(11,9)=SK(9,5)
      SK(9,11)=SK(9,5)
      SK(5,3)=-SK(9,5)
      SK(3,5)=-SK(9,5)
      SK(11,3)=-SK(9,5)
```

```
          SK(3,11)=-SK(9,5)
          RAL=RHO*A*XL
          SM(1,1)=1./3.
          SM(7,7)=SM(1,1)
          SM(7,1)=1./6.
          SM(2,2)=13./35.+6.*SMAZ/(5.*A*XL**2)
          SM(8,8)=SM(2,2)
          SM(6,2)=11.*XL/210.+SMAZ/(10.*A*XL)
          SM(8,2)=9./70.-6.*SMAZ/(5.*A*XL**2)
          SM(12,2)=-13.*XL/420.+SMAZ/(10.*A*XL)
          SM(3,3)=13./35.+6.*SMAY/(5.*A*XL**2)
          SM(9,9)=SM(3,3)
          SM(5,3)=-11.*XL/210.-SMAY/(10.*A*XL)
          SM(9,3)=9./70.-6.*SMAY/(5.*A*XL**2)
          SM(11,3)=13.*XL/420.-SMAY/(10.*A*XL)
          SM(4,4)=XPOL/(3.*A)
          SM(10,10)=SM(4,4)
          SM(10,4)=XPOL/(6.*A)
          SM(5,5)=XL**2/105.+2.*SMAY/(15.*A)
          SM(11,11)=SM(5,5)
          SM(9,5)=-13.*XL/420.+SMAY/(10.*A*XL)
          SM(11,5)=-XL*XL/140.-SMAY/(30.*A)
          SM(6,6)=XL**2/105.+2.*SMAZ/(15.*A)
          SM(12,12)=SM(6,6)
          SM(8,6)=13.*XL/420.-SMAZ/(10.*A*XL)
          SM(12,6)=-XL*XL/140.-SMAZ/(30.*A)
          SM(12,8)=-SM(6,2)
          SM(11,9)=-SM(5,3)
          DO 10 I=1,12
          DO 10 J=I,12
          IF(I.EQ.J) GO TO 1
          SM(I,J)=SM(J,I)
1         CONTINUE
10        CONTINUE
          DO 20 I=1,12
          DO 20 J=1,12
          SM(I,J)=RAL*SM(I,J)
20        CONTINUE
          RETURN
          END
C
          SUBROUTINE DIRCOS(DC,XI,XJ,XK,YI,YJ,YK,ZI,ZJ,ZK)
          DIMENSION DC(12,12),SY21(3,1),SY31(3,1),SY34(3,1),ZETAT
          (3,1),
        9 ZETA1(1,3),ZETA2(1,3),ZETA3(1,3),
        8 ZETAP(3,3),ZETAN(3,3),UNIT(3,3)
```

```
      XL=SQRT((XJ-XI)**2+(YJ-YI)**2+(ZJ-ZI)**2)
      SY21(1,1)=XJ-XI
      SY21(2,1)=YJ-YI
      SY21(3,1)=ZJ-ZI
      SY31(1,1)=XK-XI
      SY31(2,1)=YK-YI
      SY31(3,1)=ZK-ZI
      ZETA1(1,1)=SY21(1,1)/XL
      ZETA1(1,2)=SY21(2,1)/XL
      ZETA1(1,3)=SY21(3,1)/XL
      CALL UNITMX(UNIT,3)
      CALL MXTRAN(ZETAT,ZETA1,1,3)
      CALL MXPROD(ZETAP,ZETAT,ZETA1,3,1,3)
      CALL MXNEG(ZETAN,ZETAP,3,3)
      CALL MXSUM(ZETAP,UNIT,ZETAN,3,3)
      CALL MXPROD(SY34,ZETAP,SY31,3,3,1)
      XL21=SQRT(SY34(1,1)**2+SY34(2,1)**2+SY34(3,1)**2)
      ZETA2(1,1)=SY34(1,1)/XL21
      ZETA2(1,2)=SY34(2,1)/XL21
      ZETA2(1,3)=SY34(3,1)/XL21
      AXY=(XI*(YJ-YK)-YI*(XJ-XK)
     9 +(XJ*YK-YJ*XK))/2.0
      AYZ=(YI*(ZJ-ZK)-ZI*(YJ-YK)
     8 +(YJ*ZK-ZJ*YK))/2.0
      AZX=(ZI*(XJ-XK)-XI*(ZJ-ZK)
     7 +(ZJ*XK-XJ*ZK))/2.0
      TRIANG=SQRT(AYZ**2+AZX**2+AXY**2)
      ZETA3(1,1)=AYZ/TRIANG
      ZETA3(1,2)=AZX/TRIANG
      ZETA3(1,3)=AXY/TRIANG
      DO 140 II=1,3
      DC(1,II)=ZETA1(1,II)
      DC(2,II)=ZETA2(1,II)
      DC(3,II)=ZETA3(1,II)
  140 CONTINUE
      DO 125 II=1,3
      MM=1+3*II
      MN=MM+1
      N2=MM+2
      DC(MM,MM)=DC(1,1)
      DC(MM,MN)=DC(1,2)
      DC(MM,N2)=DC(1,3)
      DC(MN,MM)=DC(2,1)
      DC(MN,MN)=DC(2,2)
      DC(MN,N2)=DC(2,3)
      DC(N2,MM)=DC(3,1)
```

```
         DC(N2,MN)=DC(3,2)
         DC(N2,N2)=DC(3,3)
  125    CONTINUE
         RETURN
         END
C
C
         SUBROUTINE ASSKM(STIFF,XMS,SK,SM,I,J,NJ6,
        1ISUP,JSUP,IX,IY,IZ,IAX,IAY,IAZ,
        2JX,JY,JZ,JAX,JAY,JAZ,IDS,IMAX)
         DIMENSION STIFF(NJ6,NJ6),XMS(NJ6,NJ6),
        1SK(12,12),SM(12,12)
         DO 1 II=1,2
         IF (II.EQ.1) MM=6*I-6
         IF (II.EQ.2) MM=6*J-6
         IF (MM.LT.0) GO TO 2
         DO 3 JJ=1,2
         IF (JJ.EQ.1) MN=6*I-6
         IF (JJ.EQ.2) MN=6*J-6
         IF (MN.LT.0) GO TO 4
         NM=6*II-6
         K1=MM-IDS
         DO 5 IP1=1,6
         K1=K1+1
         L1=MN-IDS
         NM=NM+1
         N2=6*JJ-6
         DO 5 JP1=1,6
         L1=L1+1
         N2=N2+1
         STIFF(K1,L1)=STIFF(K1,L1)+SK(NM,N2)
         XMS(K1,L1)=XMS(K1,L1)+SM(NM,N2)
  5      CONTINUE
  4      CONTINUE
  3      CONTINUE
  2      CONTINUE
  1      CONTINUE
         IF (I.GT.IMAX) IMAX=I
         IF (J.GT.IMAX) IMAX=J
         NINST=6*IMAX-IDS
         IF (ISUP.GE.0) GO TO 6
         IF (IX.LT.1) GO TO 7
         NS=6*I-IDS-5
         CALL REDUCE6(NS,IDS,NINST,STIFF,XMS,NJ6)
  7      IF (IY.LT.1) GO TO 8
         NS=6*I-IDS-4
```

```
        CALL REDUCE6(NS,IDS,NINST,STIFF,XMS,NJ6)
8       IF (IZ.LT.1) GO TO 9
        NS=6*I-IDS-3
        CALL REDUCE6(NS,IDS,NINST,STIFF,XMS,NJ6)
9       IF (IAX.LT.1) GO TO 10
        NS=6*I-IDS-2
        CALL REDUCE6(NS,IDS,NINST,STIFF,XMS,NJ6)
10      IF (IAY.LT.1) GO TO 11
        NS=6*I-IDS-1
        CALL REDUCE6(NS,IDS,NINST,STIFF,XMS,NJ6)
11      IF (IAZ.LT.1)GO TO 12
        NS=6*I-IDS
        CALL REDUCE6(NS,IDS,NINST,STIFF,XMS,NJ6)
12      CONTINUE
6       CONTINUE
        IF (JSUP.GE.0) GO TO 13
        IF (JX.LT.1) GO TO 14
        NS=6*J-IDS-5
        CALL REDUCE6(NS,IDS,NINST,STIFF,XMS,NJ6)
14      IF (JY.LT.1) GO TO 15
        NS=6*J-IDS-4
        CALL REDUCE6(NS,IDS,NINST,STIFF,XMS,NJ6)
15      IF (JZ.LT.1) GO TO 16
        NS=6*J-IDS-3
        CALL REDUCE6(NS,IDS,NINST,STIFF,XMS,NJ6)
16     ·IF (JAX.LT.1) GO TO 17
        NS=6*J-IDS-2
        CALL REDUCE6(NS,IDS,NINST,STIFF,XMS,NJ6)
17      IF (JAY.LT.1) GO TO 18
        NS=6*J-IDS-1
        CALL REDUCE6(NS,IDS,NINST,STIFF,XMS,NJ6)
18      IF (JAZ.LT.1) GO TO 19
        NS=6*J-IDS
        CALL REDUCE6(NS,IDS,NINST,STIFF,XMS,NJ6)
19      CONTINUE
13      CONTINUE
        RETURN
        END
C
C
C
        SUBROUTINE REDUCE6(NS,IDS,NINST,STIFF,XMS,NJ6)
        DIMENSION STIFF(NJ6,NJ6),XMS(NJ6,NJ6)
        DO 1 I=1,NINST
        IF (I.EQ.NS) GO TO 6
        DO 5 J=1,NINST
```

```
         IF (J.EQ.NS) GO TO 7
         STIFF(I,J)=STIFF(I,J)-STIFF(I,NS)*STIFF(NS,J)/
        1STIFF(NS,NS)
         XMS(I,J)=XMS(I,J)-XMS(I,NS)*STIFF(NS,J)/
        2STIFF(NS,NS)-XMS(NS,J)*STIFF(I,NS)/STIFF(NS,NS)
        3+XMS(NS,NS)*STIFF(NS,J)*STIFF(I,NS)/(STIFF(NS,NS)**2)
7        CONTINUE
5        CONTINUE
6        CONTINUE
1        CONTINUE
         NINST=NINST-1
         IDS=IDS+1
         NS1=NS-1
         DO 2 I=1,NS1
         DO 2 J=NS,NINST
         IP1=I+1
         JP1=J+1
         STIFF(I,J)=STIFF(I,JP1)
         STIFF(J,I)=STIFF(JP1,I)
         XMS(I,J)=XMS(I,JP1)
         XMS(J,I)=XMS(JP1,I)
2        CONTINUE
         DO 3 I=NS,NINST
         DO 3 J=NS,NINST
         IP1=I+1
         JP1=J+1
         STIFF(I,J)=STIFF(IP1,JP1)
         XMS(I,J)=XMS(IP1,JP1)
3        CONTINUE
         NINST1=NINST+1
         DO 4 I=1,NINST1
         STIFF(I,NINST1)=0.0
         XMS(I,NINST1)=0.0
         STIFF(NINST1,I)=0.0
         XMS(NINST1,I)=0.0
4        CONTINUE
         RETURN
         END
C
         SUBROUTINE ELIM6(I,ISUP,IX,IY,IZ,IAX,IAY,IAZ)
C        ELIMINATES ALL DISPLACEMENTS
         ISUP=-100
         I=I/1000
         IX=1
         IY=1
         IZ=1
```

```
         IAX=1
         IAY=1
         IAZ=1
         RETURN
         END
C
         SUBROUTINE ISOLATE(I,ISUP,IX,IY,IZ,IAX,IAY,IAZ)
C        ELIMINATES ALL BUT U AND V DISPLACEMENTS
         ISUP=I
         I=-I
         IX=0
         IY=0
         IZ=0
         IAX=1
         IAY=1
         IAZ=1
         RETURN
         END
C
         SUBROUTINE EIGPT(ST,N,EV,VEC,NV)
         DIMENSION ST(N,N),EV(NV),VEC(N,NV)
         DO 53 I=1,NV
         WRITE(6,120) EV(I)
  120    FORMAT(///12H EIGENVALUE=,E15.5)
         EV(I)=SQRT(1.0/EV(I))/(2.0*3.1415927)
         WRITE(6,121) EV(I)
  121    FORMAT(//11H FREQUENCY=,E15.5)
         DO 53 J=1,N
         WRITE(6,122) VEC(J,I)
  122    FORMAT(10X,E15.5)
   53    CONTINUE
         RETURN
         END
C
C
         SUBROUTINE UNITMX(A,N)
         DIMENSION A(1)
C        (A)=(I)
         DO 1 I=1,N
         DO 1 J=1,N
         A((J-1)*N+I)=0.
         A((I-1)*N+I)=1.
    1    CONTINUE
         RETURN
         END
C
```

```
      SUBROUTINE NULL(A,M,N)
      DIMENSION A(M,N)
      DO 1 I=1,M
      DO 1 J=1,N
      A(I,J)=0.0
1     CONTINUE
      RETURN
      END
      SUBROUTINE MXPROD(AB,A,B,N,M,L)
      DIMENSION AB(N,L),A(N,M),B(M,L)
C     (AB)=(A)*(B)
      DO 1 I=1,N
      DO 1 J=1,L
      AB(I,J)=0.0
1     CONTINUE
      DO 2 I=1,N
      DO 2 K=1,L
      DO 2 J=1,M
      AB(I,K)=AB(I,K)+A(I,J)*B(J,K)
2     CONTINUE
      RETURN
      END
      SUBROUTINE VALEIG(A,Y,X2,N,D,N1)
      DIMENSION A(1),Y(1),
     2Z(200),YO(200)
      X1=100000.0
      DO 1 I=1,N
      Y(I)=1.0
1     CONTINUE
      X2=-100000.0
2     CONTINUE
      DO 20 I=1,N
      YO(I)=Y(I)
20    CONTINUE
      DO 4 I=1,N
      SUM=0.0
          DO 5  J=1,N
          SUM=SUM+A((J-1)*N1+I)*Y(J)
5         CONTINUE
      Z(I)=SUM
4       CONTINUE
      X2=0.0
      DO 21 I=1,N
      IF(ABS(X2).LT.ABS(Z(I)))  X2=Z(I)
21    CONTINUE
      DO 22 I=1,N
```

```
         Y(I)=Z(I)/X2
22       CONTINUE
         IF(ABS((X1-X2)/X2).GT.D) GO TO  6
         GO TO 66
6        CONTINUE
         X1=X2
           GO TO  2
66       CONTINUE
         X = 0.0
         DO 11 I=1,N
        IF(ABS(X).LT.ABS(Y(I))) X=Y(I)
11       CONTINUE
         DO 12 I=1,N
         Y(I)=Y(I)/X
12       CONTINUE
         RETURN
         END
C
         SUBROUTINE EIGEN(A,N,AM,VEC,M1,D,XX,NP1)
         DIMENSION A(1),
        1 VEC(1),AM(1),XX(1),
        2 X(200),IG(200)
          MN=N
101       FORMAT(E20.8)
         NN = N
         CALL VALEIG(A,X,XM,NN,D,N)
          M=1
          DO 2 I=1,NN
         I3=(M-1)*N+I
          VEC(I3)=X(I)
          XX(I3)=VEC(I3)
2        CONTINUE
         AM(M) = XM
         IF(M1.LT.2) RETURN
         DO 20 M=2,M1
          DO 4 I=1,NN
          IF (ABS(XX((M-2)*N+I)-1.0).LT.0.0001)IR=I
4        CONTINUE
          IG(M-1)=IR
         DO 5 I=1,NN
          XX((MN-M+2)*N+MN-I+1)=A((I-1)*N+IR)
5        CONTINUE
          DO 6 I=1,NN
          DO 6 J=1,NN
         I3=(J-1)*N+I
         I4=(M-2)*N+I
```

```
        I5=(MN-M+2)*N+MN-J+1
        A(I3)=A(I3)-XX(I4)*XX(I5)
6       CONTINUE
        DO 9 I=1,NN
        IF(I.EQ.IR) GO TO 8
        IF(I.GT.IR) K1=I-1
        IF(I.LE.IR) K1=I
            DO 7 J=1,NN
            IF(J.EQ.IR) GO TO 7
            IF(J.GT.IR) K2=J-1
        IF(J.LE.IR) K2=J
            A((K2-1)*N+K1) = A((J-1)*N+I)
7           CONTINUE
8     CONTINUE
9       CONTINUE
      NN = NN-1
      M3=NN
      IF (M.NE.MN)  GO TO  11
      XM=A(1)
      X(1)=1.0
      GO TO  12
11    CALL VALEIG(A,X,XM,NN,D,N)
      WRITE(6,101)XM
12      DO 13 I=1,NN
      XX((M-1)*N+I)=X(I)
13      CONTINUE
      AM(M)=XM
      M11 = M-1
      MM1=1000-M11
      DO 20 M8=MM1,999
       M12 = M3+1
      M2=1000-M8
       M22 =IG(M2)+1
      MM22=1000-M22
      MM12=1000-M12
      IF(M12.LT.M22)GO TO 14
      DO 15 I3=MM12,MM22
      I=1000-I3
      X(I)=X(I-1)
15    CONTINUE
14      CONTINUE
      J=IG(M2)
      X(J)=0.0
      SUM=0.0
      DO 16 I= 1,M12
      SUM=SUM+XX((MN-M2+2-1)*N+MN-I+1)*X(I)
```

```
16      CONTINUE
        XK=(AM(M2)-XM)/SUM
        DO 17 I= 1,M12
        X(I)=XX((M2-1)*N+I)-XK*X(I)
17      CONTINUE
        SUM=0.0
        DO 18 I=1,M12
        IF (ABS(SUM).LT.ABS(X(I)))SUM=X(I)
18      CONTINUE
        DO 19 I=1,M12
        X(I)=X(I)/SUM
19      CONTINUE
        M3=M3+1
        IF (M2.NE.1)   GO TO 20
        DO 3 I=1,M3
        VEC((M-1)*N+I)=X(I)
3       CONTINUE
20      CONTINUE
        RETURN
        END
        SUBROUTINE INVMX(A,N,D,L,M,B)
        DIMENSION A(1),B(1),L(1),M(1)
        I5=0
3       I1=N-1
        DO 1 I=1,I1
        I2=I+1
        DO 2 J=I2,N
        I3=(I-1)*N+J
        I4=(J-1)*N+I
        B(J)=A(I3)
        A(I3)=A(I4)
2       A(I4)=B(J)
1       CONTINUE
        IF (I5.EQ.1) RETURN
        NK=-N
        DO 80 K=1,N
        NK=NK+N
        L(K)=K
        M(K)=K
        KK=NK+K
        BIGA=A(KK)
        DO 20 J=K,N
        IZ=N*(J-1)
        DO 20 I=K,N
        IJ=IZ+I
10      IF (ABS(BIGA)-ABS(A(IJ))) 15,20,20
```

```
15    BIGA=A(IJ)
      L(K)=I
      M(K)=J
20    CONTINUE
      J=L(K)
      IF (J-K) 35,35,25
25    KI=K-N
      DO 30 I=1,N
      KI=KI+N
      HOLD=-A(KI)
      JI=KI-K+J
      A(KI)=A(JI)
30    A(JI)=HOLD
35    I=M(K)
      IF (I-K) 45,45,38
38    JP=N*(I-1)
      DO 40 J=1,N
      JK=NK+J
      JI=JP+J
      HOLD=-A(JK)
      A(JK)=A(JI)
40    A(JI)=HOLD
45    IF (BIGA) 48,46,48
46    D=0.0
      RETURN
48    DO 55 I=1,N
      IF (I-K) 50,55,50
50    IK=NK+I
      A(IK)=A(IK)/(-BIGA)
55    CONTINUE
      DO 65 I=1,N
      IK=NK+I
      HOLD=A(IK)
      IJ=I-N
      DO 65 J=1,N
      IJ=IJ+N
      IF(I-K) 60,65,60
60    IF (J-K) 62,65,62
62    KJ=IJ-I+K
      A(IJ)=HOLD*A(KJ)+A(IJ)
65    CONTINUE
      KJ=K-N
      DO 75 J=1,N
      KJ=KJ+N
      IF (J-K) 70,75,70
70    A(KJ)=A(KJ)/BIGA
```

```
 75    CONTINUE
       D=D*BIGA
       A(KK)=1.0/BIGA
 80    CONTINUE
       K=N
100    K=(K-1)
       IF (K) 150,150,105
105    I=L(K)
       IF (I-K) 120,120,108
108    JQ=N*(K-1)
       JR=N*(I-1)
       DO 110 J=1,N
       JK=JQ+J
       HOLD=A(JK)
       JI=JR+J
       A(JK)=-A(JI)
110    A(JI)=HOLD
120    J=M(K)
       IF (J-K) 100,100,125
125    KI=K-N
       DO 130 I=1,N
       KI=KI+N
       HOLD=A(KI)
       JI=KI-K+J
       A(KI)=-A(JI)
130    A(JI)=HOLD
       GO TO 100
150    I5=1
       GO TO 3
       END
       SUBROUTINE PRODAB(A,B,N)
       DIMENSION A(N,N),B(N,N),VEC(100,1)
C      (A)=(A)*(B)
       DO 1 I=1,N
       DO 2 J=1,N
       VEC(J,1)=0.0
       DO 2 K=1,N
       VEC(J,1)=VEC(J,1)+A(I,K)*B(K,J)
  2    CONTINUE
       DO 1 J=1,N
       A(I,J)=VEC(J,1)
  1    CONTINUE
       RETURN
       END
       SUBROUTINE MXTRAN(AT,A,M,N)
       DIMENSION AT(N,M),A(M,N)
```

```
C      (AT)=(A)T
       DO 1 I=1,M
       DO 1 J=1,N
       AT(J,I)=A(I,J)
   1   CONTINUE
       RETURN
       END
       SUBROUTINE MXNEG(A,B,M,N)
       DIMENSION A(M,N),B(M,N)
C      (A)=-(B)
       DO 1 I=1,M
       DO 1 J=1,N
       A(I,J)=-B(I,J)
   1   CONTINUE
       RETURN
       END
       SUBROUTINE MXSUM(A,B,C,M,N)
       DIMENSION A(M,N),B(M,N),C(M,N)
C      (A)=(B)+(C)
       DO 1 I=1,M
       DO 1 J=1,N
       A(I,J)=B(I,J)+C(I,J)
   1   CONTINUE
       RETURN
       END
```

A 5.2 A Typical Input File for VIBDAT3D.DAT (Example 8.3)

```
4    4    0    0    12   6    2    8
1E-2    1E-2    20.01
10.01   1E-2    19.99
10.01   10.01   20.01
1E-2    10.01   19.99
7.35E-4    3E7    1.15E7    6.3E-2    1.3E-3    3.26E-4    3.26E-4    6.51E-4
7.35E-4    3E7    1.15E7    .126    2.6E-3    6.52E-4    6.52E-4    1.302E-3
0    1    4
2
0    2    3
2
0    3    2
2
0    4    1
2
```

```
1   2   3
1
-1   4   3
1
-2   3   4
1
-3  -4   1
1
2
1   5E-4
7   1E-3
0   0   0
10   0   0
10  10   0
0  10   0
```

A 5.3 A Typical Output File for OUT11.OUT (Example 8.3)

BASIC PARAMETERS

```
        NUMBER OF ACTUAL (NON-ZERO) NODAL POINTS . . .    4
        NUMBER OF FIXED (ZERO) NODES . . . .    4
        NO. OF ZERO DISPLTS. AT PIN-JOINTS, ETC. . . . . .    0
        NUMBER OF IMAGINARY NODES. .    0
        NUMBER OF FREE DISPLACEMENTS    12
        NUMBER OF FREQUENCIES . .    6
        NUMBER OF MEMBER TYPES . .    2
        NUMBER OF MEMBERS   . . . . .    8
```

          COORDINATES

```
    XI          YI          ZI
 0.10000E-01 0.10000E-01 0.20010E+02
 0.10010E+02 0.10000E-01 0.19990E+02
 0.10010E+02 0.10010E+02 0.20010E+02
 0.10000E-01 0.10010E+02 0.19990E+02
```

```
    RHO          E          G          AREA          TC
 IY          IZ          JP
  0.73500E-03 0.30000E+08 0.11500E+08 0.63000E-01 0.13000E-02
 0.32600E-03 0.32600E-03 0.65100E-03
```

```
 0.73500E-03 0.30000E+08 0.11500E+08 0.12600E+00 0.26000E-02
0.65200E-03 0.65200E-03 0.13020E-02
```

```
            I          J          K
            0          1          4
MATERIAL TYPE=    2
            0          2          3
MATERIAL TYPE=    2
            0          3          2
MATERIAL TYPE=    2
            0          4          1
MATERIAL TYPE=    2
            1          2          3
MATERIAL TYPE=    1
           -1          4          3
MATERIAL TYPE=    1
           -2          3          4
MATERIAL TYPE=    1
           -3         -4          1
MATERIAL TYPE=    1
NO. OF NODES WITH CONCENTRATED MASSES=    2
THE FINISHED U DISPT. POSITION OF MASS ITS VALUE ARE    1
        0.50000000E-03
THE FINISHED U DISPT. POSITION OF MASS ITS VALUE ARE    7
        0.10000000E-02
 0.00000E+00 0.00000E+00 0.00000E+00
 0.10000E+02 0.00000E+00 0.00000E+00
 0.10000E+02 0.10000E+02 0.00000E+00
 0.00000E+00 0.10000E+02 0.00000E+00
           6
        0.69258838E-04
        0.25905472E-04
        0.50910663E-05
        0.10604264E-07
        0.79734927E-08
```

```
EIGENVALUE=    0.69008E-04
```

```
FREQUENCY=    0.19159E+02
              0.98427E+00
              0.98359E+00
             -0.64784E-03
```

```
            0.98425E+00
            0.99813E+00
           -0.98988E-03
            0.10000E+01
            0.99818E+00
           -0.13383E-02
            0.99996E+00
            0.98357E+00
           -0.99001E-03
```

EIGENVALUE=   0.69259E-04


```
FREQUENCY=      0.19124E+02
                0.90453E+00
               -0.56997E+00
               -0.11013E-03
                0.90451E+00
               -0.66016E+00
               -0.38859E-03
                0.10000E+01
               -0.66019E+00
               -0.22760E-03
                0.99996E+00
               -0.56996E+00
                0.51946E-04
```


EIGENVALUE=   0.25905E-04


```
FREQUENCY=      0.31270E+02
                0.10000E+01
               -0.99835E+00
               -0.54095E-06
                0.99995E+00
                0.75387E+00
               -0.91240E-03
               -0.75216E+00
                0.75395E+00
               -0.11758E-05
               -0.75214E+00
               -0.99829E+00
                0.91067E-03
```

```
EIGENVALUE=     0.50911E-05


FREQUENCY=      0.70537E+02
                0.99935E+00
                0.99998E+00
               -0.10585E-02
                0.99909E+00
               -0.82491E+00
               -0.35327E-04
               -0.82534E+00
               -0.82533E+00
                0.77946E-03
               -0.82537E+00
                0.10000E+01
               -0.35741E-04




EIGENVALUE=     0.10604E-07


FREQUENCY=      0.15455E+04
               -0.23154E-02
               -0.26114E-02
                0.18429E+00
                0.10505E-01
               -0.46048E-01
                0.74013E-01
               -0.50821E-01
               -0.62609E-01
                0.10000E+01
               -0.68294E-01
                0.85787E-02
                0.78738E-01




EIGENVALUE=     0.79735E-08


FREQUENCY=      0.17824E+04
                0.67738E-01
                0.68031E-01
                0.10000E+01
                0.68551E-01
```

```
                    0.32552E-02
                    0.10992E+00
                    0.17790E-01
                    0.14480E-01
                    0.69423E-02
                   -0.19025E-02
                    0.66411E-01
                    0.11689E+00
```

A 5.4 Input File for Example 8.4

```
     44    4    0    0    42    3    3    124
   1.25001        1.25001        29.7501
   4.125      1.25       29.75
   4.125      4.125      29.75
    1.25      4.125      29.75
   2.688      1.146      26.938
   4.229      2.688      26.938
   2.688      4.229      26.938
   1.146      2.688      26.938
   1.027      1.027      24.125
   4.348      1.027      24.125
   4.348      4.348      24.125
   1.027      4.348      24.125
   2.688      0.915      21.5
   4.46       2.688      21.5
   2.688      4.460      21.5
   0.915      2.688      21.5
   0.803      0.803      18.875
   4.572      0.803      18.875
   4.572      4.572      18.875
   0.803      4.572      18.875
   2.688      0.68       16.0
   4.694      2.688      16.0
  2.688    4.694       16.0
  0.68     2.688       16.0
  0.559    0.559       13.125
  4.816    0.559       13.125
  4.816    4.816       13.125
  0.559    4.816       13.125
  2.688    0.426       10.0
  4.95     2.688       10.0
  2.688    4.95        10.0
```

```
     0.426      2.688      10.0
     0.293      0.293      6.875
     5.082      0.293      6.875
     5.082      5.082      6.875
     0.293      5.082      6.875
     2.688      0.178      4.188
     5.197      2.688      4.188
     2.688      5.197      4.188
     0.178      2.688      4.188
     0.032      0.032      0.75
     5.343      0.032      0.75
     5.343      5.343      0.75
     0.032      5.343      0.75
 0.00079  14000000.   5190000.    0.0103   0.00008746   0.00004373
0.00004373   8.746E-5
 0.00079  14000000.   5190000.    0.0159    0.0002854    0.0001427
0.0001427   2.854E-4
 0.875    14000000.   5190000.    0.0103   0.00008746   0.00004373
0.00004373   8.746E-5
              1           2           3
              3
              1           4           3
              3
              2           3           4
              3
              3           4           1
              3
              1           9          12
              2
              1           5           9
              1
             -1           8           4
              1
              2          10          11
              2
              2           6          10
              1
             -2           5           1
              1
              3           7          11
              1
              3           6          11
              1
             -3          11          12
              2
              4          12           9
```

2		
4	8	1
1		
-4	7	3
1		
5	10	9
1		
5000	9	10
1		
6	11	10
1		
6000	10	11
1		
7	11	12
1		
7000	12	11
1		
8	12	9
1		
8000	9	12
1		
9	12	11
1		
9	10	11
1		
9	17	18
2		
9	13	17
1		
-9	16	12
1		
10	11	12
1		
10	18	11
2		
10	14	11
1		
-10	13	9
1		
11	12	20
1		
11	14	10
1		
11	15	12
1		
12	20	17

2		
-11	19	18
2		
12	15	11
1		
-12	16	9
1		
13	17	18
1		
13000	18	17
1		
14	18	19
1		
14000	19	18
1		
15	20	19
1		
15000	19	20
1		
16	17	20
1		
16000	20	17
1		
17	25	26
2		
17	18	26
1		
17	20	28
1		
17	21	25
1		
-17	24	25
1		
18	19	20
1		
18	26	27
2		
18	22	19
1		
18000	21	17
1		
19	27	28
2		
19	20	28
1		
19	22	18

1		
-19	23	20
1		
20	28	27
2		
20	23	19
1		
20000	24	17
1		
21	25	26
1		
21000	26	25
1		
22	26	27
1		
22000	27	26
1		
23	27	28
1		
23000	28	27
1		
24	28	25
1		
24000	25	28
1		
25	33	34
2		
25	26	27
1		
25	28	27
1		
25	29	33
1		
25000	32	28
1		
26	34	35
2		
26	27	35
1		
26	30	27
1		
-26	29	25
1		
27	35	36
2		
27	28	36

1		
27	30	26
1		
27000	31	28
1		
28	36	35
2		
28	31	27
1		
−28	32	25
1		
29	34	33
1		
29000	33	34
1		
30	35	34
1		
30000	34	35
1		
31	35	36
1		
31000	36	35
1		
32	36	33
1		
32000	33	36
1		
33	41	42
2		
33	34	35
1		
33	36	35
1		
33	37	41
1		
−33	40	36
1		
34	35	43
1		
34	42	43
2		
34	37	42
1		
34000	38	35
1		
35	43	44

```
            2
          35          36          44
            1
          35          38          34
            1
         -35          39          43
            1
          36          44          43
            2
          36          39          35
            1
       36000          40          33
            1
          37          42          41
            1
       37000          41          42
            1
          38          43          42
            1
       38000          42          43
            1
          39          43          44
            1
       39000          44          43
            1
          40          44          41
            1
       40000          41          44
            1
       41000           0           0
            2
       42000           0           0
            2
       43000           0           0
            2
       44000           0           0
            2
0
0.    0.    0.
5.375  0.    0.
5.375  0.    0.
0.    0.    0.
5.375  5.375  0.
5.375  0.    0.
0.    5.375  0.
0.    0.    0.
```

A 5.5 Output File for Example 8.4

```
***BASIC PARAMETERS***

        NUMBER OF ACTUAL (NON-ZERO) NODAL POINTS . . .    44
        NUMBER OF FIXED (ZERO) NODES . . . .    4
        NO. OF ZERO DISPLTS. AT PIN-JOINTS, ETC. . . . . .    0
        NUMBER OF IMAGINARY NODES. .    0
        NUMBER OF FREE DISPLACEMENTS    42
        NUMBER OF FREQUENCIES . .     3
        NUMBER OF MEMBER TYPES . .    3
        NUMBER OF MEMBERS   . . . . .  124
```

```
        COORDINATES

      XI          YI          ZI
  0.12500E+01  0.12500E+01  0.29750E+02
  0.41250E+01  0.12500E+01  0.29750E+02
  0.41250E+01  0.41250E+01  0.29750E+02
  0.12500E+01  0.41250E+01  0.29750E+02
  0.26880E+01  0.11460E+01  0.26938E+02
  0.42290E+01  0.26880E+01  0.26938E+02
  0.26880E+01  0.42290E+01  0.26938E+02
  0.11460E+01  0.26880E+01  0.26938E+02
  0.10270E+01  0.10270E+01  0.24125E+02
  0.43480E+01  0.10270E+01  0.24125E+02
  0.43480E+01  0.43480E+01  0.24125E+02
  0.10270E+01  0.43480E+01  0.24125E+02
  0.26880E+01  0.91500E+00  0.21500E+02
  0.44600E+01  0.26880E+01  0.21500E+02
  0.26880E+01  0.44600E+01  0.21500E+02
  0.91500E+00  0.26880E+01  0.21500E+02
  0.80300E+00  0.80300E+00  0.18875E+02
  0.45720E+01  0.80300E+00  0.18875E+02
  0.45720E+01  0.45720E+01  0.18875E+02
  0.80300E+00  0.45720E+01  0.18875E+02
  0.26880E+01  0.68000E+00  0.16000E+02
  0.46940E+01  0.26880E+01  0.16000E+02
  0.26880E+01  0.46940E+01  0.16000E+02
  0.68000E+00  0.26880E+01  0.16000E+02
  0.55900E+00  0.55900E+00  0.13125E+02
  0.48160E+01  0.55900E+00  0.13125E+02
  0.48160E+01  0.48160E+01  0.13125E+02
```

```
0.55900E+00 0.48160E+01 0.13125E+02
0.26880E+01 0.42600E+00 0.10000E+02
0.49500E+01 0.26880E+01 0.10000E+02
0.26880E+01 0.49500E+01 0.10000E+02
0.42600E+00 0.26880E+01 0.10000E+02
0.29300E+00 0.29300E+00 0.68750E+01
0.50820E+01 0.29300E+00 0.68750E+01
0.50820E+01 0.50820E+01 0.68750E+01
0.29300E+00 0.50820E+01 0.68750E+01
0.26880E+01 0.17800E+00 0.41880E+01
0.51970E+01 0.26880E+01 0.41880E+01
0.26880E+01 0.51970E+01 0.41880E+01
0.17800E+00 0.26880E+01 0.41880E+01
0.32000E-01 0.32000E-01 0.75000E+00
0.53430E+01 0.32000E-01 0.75000E+00
0.53430E+01 0.53430E+01 0.75000E+00
0.32000E-01 0.53430E+01 0.75000E+00
```

```
      RHO          E           G          AREA         TC
IY            IZ           JP
 0.79000E-03 0.14000E+08 0.51900E+07 0.10300E-01 0.87460E-04
0.43730E-04 0.43730E-04
 0.87460E-04
 0.79000E-03 0.14000E+08 0.51900E+07 0.15900E-01 0.28540E-03
0.14270E-03 0.14270E-03
 0.28540E-03
 0.87500E+00 0.14000E+08 0.51900E+07 0.10300E-01 0.87460E-04
0.43730E-04 0.43730E-04
 0.87460E-04
```

```
            I          J          K
            1          2          3
MATERIAL TYPE=      3
            1          4          3
MATERIAL TYPE=      3
            2          3          4
MATERIAL TYPE=      3
            3          4          1
MATERIAL TYPE=      3
            1          9         12
MATERIAL TYPE=      2
            1          5          9
MATERIAL TYPE=      1
           -1          8          4
```

```
MATERIAL TYPE=        1
               2       10        11
MATERIAL TYPE=        2
               2        6        10
MATERIAL TYPE=        1
              -2        5         1
MATERIAL TYPE=        1
               3        7        11
MATERIAL TYPE=        1
               3        6        11
MATERIAL TYPE=        1
              -3       11        12
MATERIAL TYPE=        2
               4       12         9
MATERIAL TYPE=        2
               4        8         1
MATERIAL TYPE=        1
              -4        7         3
MATERIAL TYPE=        1
               5       10         9
MATERIAL TYPE=        1
            5000        9        10
MATERIAL TYPE=        1
               6       11        10
MATERIAL TYPE=        1
            6000       10        11
MATERIAL TYPE=        1
               7       11        12
MATERIAL TYPE=        1
            7000       12        11
MATERIAL TYPE=        1
               8       12         9
MATERIAL TYPE=        1
            8000        9        12
MATERIAL TYPE=        1
               9       12        11
MATERIAL TYPE=        1
               9       10        11
MATERIAL TYPE=        1
               9       17        18
MATERIAL TYPE=        2
               9       13        17
MATERIAL TYPE=        1
              -9       16        12
MATERIAL TYPE=        1
              10       11        12
```

```
MATERIAL TYPE=      1
         10         18          11
MATERIAL TYPE=      2
         10         14          11
MATERIAL TYPE=      1
        -10         13           9
MATERIAL TYPE=      1
         11         12          20
MATERIAL TYPE=      1
         11         14          10
MATERIAL TYPE=      1
         11         15          12
MATERIAL TYPE=      1
         12         20          17
MATERIAL TYPE=      2
        -11         19          18
MATERIAL TYPE=      2
         12         15          11
MATERIAL TYPE=      1
        -12         16           9
MATERIAL TYPE=      1
         13         17          18
MATERIAL TYPE=      1
      13000         18          17
MATERIAL TYPE=      1
         14         18          19
MATERIAL TYPE=      1
      14000         19          18
MATERIAL TYPE=      1
         15         20          19
MATERIAL TYPE=      1
      15000         19          20
MATERIAL TYPE=      1
         16         17          20
MATERIAL TYPE=      1
      16000         20          17
MATERIAL TYPE=      1
         17         25          26
MATERIAL TYPE=      2
         17         18          26
MATERIAL TYPE=      1
         17         20          28
MATERIAL TYPE=      1
         17         21          25
MATERIAL TYPE=      1
        -17         24          25
```

```
MATERIAL TYPE=      1
        18          19          20
MATERIAL TYPE=      1
        18          26          27
MATERIAL TYPE=      2
        18          22          19
MATERIAL TYPE=      1
     18000          21          17
MATERIAL TYPE=      1
        19          27          28
MATERIAL TYPE=      2
        19          20          28
MATERIAL TYPE=      1
        19          22          18
MATERIAL TYPE=      1
       -19          23          20
MATERIAL TYPE=      1
        20          28          27
MATERIAL TYPE=      2
        20          23          19
MATERIAL TYPE=      1
     20000          24          17
MATERIAL TYPE=      1
        21          25          26
MATERIAL TYPE=      1
     21000          26          25
MATERIAL TYPE=      1
        22          26          27
MATERIAL TYPE=      1
     22000          27          26
MATERIAL TYPE=      1
        23          27          28
MATERIAL TYPE=      1
     23000          28          27
MATERIAL TYPE=      1
        24          28          25
MATERIAL TYPE=      1
     24000          25          28
MATERIAL TYPE=      1
        25          33          34
MATERIAL TYPE=      2
        25          26          27
MATERIAL TYPE=      1
        25          28          27
MATERIAL TYPE=      1
        25          29          33
```

```
MATERIAL TYPE=    1
     25000       32        28
MATERIAL TYPE=    1
       26        34        35
MATERIAL TYPE=    2
       26        27        35
MATERIAL TYPE=    1
       26        30        27
MATERIAL TYPE=    1
      -26        29        25
MATERIAL TYPE=    1
       27        35        36
MATERIAL TYPE=    2
       27        28        36
MATERIAL TYPE=    1
       27        30        26
MATERIAL TYPE=    1
     27000       31        28
MATERIAL TYPE=    1
       28        36        35
MATERIAL TYPE=    2
       28        31        27
MATERIAL TYPE=    1
      -28        32        25
MATERIAL TYPE=    1
       29        34        33
MATERIAL TYPE=    1
     29000       33        34
MATERIAL TYPE=    1
       30        35        34
MATERIAL TYPE=    1
     30000       34        35
MATERIAL TYPE=    1
       31        35        36
MATERIAL TYPE=    1
     31000       36        35
MATERIAL TYPE=    1
       32        36        33
MATERIAL TYPE=    1
     32000       33        36
MATERIAL TYPE=    1
       33        41        42
MATERIAL TYPE=    2
       33        34        35
MATERIAL TYPE=    1
       33        36        35
```

```
MATERIAL TYPE=     1
           33     37         41
MATERIAL TYPE=     1
          -33     40         36
MATERIAL TYPE=     1
           34     35         43
MATERIAL TYPE=     1
           34     42         43
MATERIAL TYPE=     2
           34     37         42
MATERIAL TYPE=     1
        34000     38         35
MATERIAL TYPE=     1
           35     43         44
MATERIAL TYPE=     2
           35     36         44
MATERIAL TYPE=     1
           35     38         34
MATERIAL TYPE=     1
          -35     39         43
MATERIAL TYPE=     1
           36     44         43
MATERIAL TYPE=     2
           36     39         35
MATERIAL TYPE=     1
        36000     40         33
MATERIAL TYPE=     1
           37     42         41
MATERIAL TYPE=     1
        37000     41         42
MATERIAL TYPE=     1
           38     43         42
MATERIAL TYPE=     1
        38000     42         43
MATERIAL TYPE=     1
           39     43         44
MATERIAL TYPE=     1
        39000     44         43
MATERIAL TYPE=     1
           40     44         41
MATERIAL TYPE=     1
        40000     41         44
MATERIAL TYPE=     1
        41000      0          0
MATERIAL TYPE=     2
        42000      0          0
```

```
MATERIAL TYPE=      2
      43000           0           0
MATERIAL TYPE=      2
      44000           0           0
MATERIAL TYPE=      2
NO. OF NODES WITH CONCENTRATED MASSES=      0
 0.00000E+00 0.00000E+00 0.00000E+00
 0.53750E+01 0.00000E+00 0.00000E+00
 0.53750E+01 0.00000E+00 0.00000E+00
 0.00000E+00 0.00000E+00 0.00000E+00
 0.53750E+01 0.53750E+01 0.00000E+00
 0.53750E+01 0.00000E+00 0.00000E+00
 0.00000E+00 0.53750E+01 0.00000E+00
 0.00000E+00 0.00000E+00 0.00000E+00
NO. OF REDUCED DISPLACEMENTS=   222
           3
      0.16798670E-03
      0.18193796E-04



  EIGENVALUE=      0.16802E-03


  FREQUENCY=      0.12279E+02
                  0.99954E+00
                  0.10000E+01
                  0.14473E+00
                  0.99848E+00
                  0.99891E+00
                  0.41524E-04
                  0.99952E+00
                  0.99997E+00
                 -0.14470E+00
                  0.99846E+00
                  0.99894E+00
                 -0.23193E-04
                  0.69795E+00
                  0.69828E+00
                  0.15783E+00
                  0.69460E+00
                  0.69492E+00
                  0.43902E-04
                  0.69795E+00
                  0.69827E+00
                 -0.15780E+00
```

```
                          0.69460E+00
                          0.69493E+00
                         -0.27385E-04
                          0.45120E+00
                          0.45141E+00
                          0.15204E+00
                          0.45121E+00
                          0.45142E+00
                         -0.15202E+00
                          0.22433E+00
                          0.22445E+00
                          0.35331E-04
                          0.22434E+00
                          0.22444E+00
                         -0.23144E-04
                          0.76413E-01
                          0.76443E-01
                          0.78553E-01
                          0.76424E-01
                          0.76452E-01
                         -0.78549E-01
```

```
EIGENVALUE=       0.16799E-03


FREQUENCY=        0.12280E+02
                          0.99999E+00
                         -0.16135E-01
                          0.71212E-01
                          0.10000E+01
                         -0.17209E-01
                         -0.73618E-01
                          0.99998E+00
                         -0.16148E-01
                         -0.71202E-01
                          0.99999E+00
                         -0.17196E-01
                          0.73625E-01
                          0.69694E+00
                         -0.99398E-02
                          0.77657E-01
                          0.69699E+00
                         -0.13300E-01
                         -0.80282E-01
```

```
          0.69694E+00
         -0.99454E-02
         -0.77647E-01
          0.69700E+00
         -0.13294E-01
          0.80288E-01
          0.44823E+00
         -0.41160E-02
          0.74807E-01
          0.44824E+00
         -0.41168E-02
         -0.74799E-01
          0.22972E+00
         -0.88965E-02
         -0.64419E-01
          0.22972E+00
         -0.89077E-02
          0.64424E-01
          0.69146E-01
          0.60646E-02
          0.38651E-01
          0.69152E-01
          0.60681E-02
         -0.38650E-01
```

```
EIGENVALUE=     0.18194E-04
```

```
FREQUENCY=      0.37313E+02
               -0.99995E+00
               -0.99990E+00
               -0.89865E-01
               -0.99994E+00
                0.99978E+00
                0.87727E-01
                0.99960E+00
                0.10000E+01
               -0.89826E-01
                0.99960E+00
               -0.99988E+00
                0.87708E-01
               -0.50012E+00
               -0.50009E+00
               -0.96039E-01
```

```
        -0.50005E+00
         0.49995E+00
         0.94243E-01
         0.49989E+00
         0.50003E+00
        -0.96007E-01
         0.49982E+00
        -0.50001E+00
         0.94223E-01
        -0.21385E+00
        -0.21384E+00
        -0.69335E-01
         0.21369E+00
         0.21379E+00
        -0.69314E-01
        -0.57776E-01
         0.57807E-01
         0.37699E-01
         0.57752E-01
        -0.57775E-01
         0.37683E-01
         0.21382E-02
         0.21306E-02
        -0.14646E-01
        -0.21486E-02
        -0.21265E-02
        -0.14639E-01
```

Appendix 6. Computer Program for the Vibration of Cross-stiffened Grids

A 6.1 The Computer Program

```
C     VIBRATION OF FLAT GRILLAGES
      DIMENSION STIFF(50,50),XMS(50,50),
    8 VEC(50,10)
C     MODIFIED FOR IBM PC ON 21st JULY 1990
C     COPYRIGHT OF DR.C.T.F.ROSS.
      OPEN(5,FILE='GRIDIN.DAT',STATUS='OLD')
      OPEN(6,FILE='GRIDOUT.DAT',STATUS='OLD')
      READ(5,*)NJ,NFIXD,NP
      NJ3=3*NJ
      CALL NULL(STIFF,NJ3,NJ3)
      CALL NULL(XMS,NJ3,NJ3)
      CALL VBFLGR(STIFF,XMS,NJ,NFIXD,NP,NJ3,VEC)
      close(5)
      close(6)
      STOP
      END
C
      SUBROUTINE VBFLGR(STIFF,XMS,NJ,NFIXD,NP,NJ3,VEC)
      DIMENSION STIFF(NJ3,NJ3),XMS(NJ3,NJ3),
    8 DC(6,6),DCT(6,6),SK(6,6),SKX(6,6),SM(6,6),
    7 XLAM(20),VEC(NJ3,NP),
    6 LVEC(50),MVEC(50),BVEC(50),
    5 NSUPR(50),NPOSS(50),NXSUP(50),NYSUP(50),NZSUP(50),
    4 X(50),Y(50)
C
C
      READ(5,*)MEMX,MEMY
      MEMS=MEMX+MEMY
      WRITE(6,130) NJ,NFIXD,NP,MEMX,MEMY
  130 FORMAT(8X,23H ***BASIC PARAMETERS***//
    1 5X,27HNUMBER OF NODES * * * * * *,I2/
    2 5X,'HNUMBER OF NODES WITH ZERO DISPLACEMENTS=',I2/
    3 5X,27HNUMBER OF EIGEN VALUES* * *,I2/
    4 5X,27HNUMBER OF X DIRECTION MEMS*,I2/
    5 5X,27HNUMBER OF Y DIRECTION MEMS*,I2)
      WRITE(6,131)
  131 FORMAT(///2X,22H***NODAL POINT DATA***//
    1 2X,4HNODE,5X,1HX,9X,1HY)
      DO 1 I=1,NJ
```

```
          READ(5,*)X(I),Y(I)
          WRITE(6,102)I,X(I),Y(I)
1         CONTINUE
102       FORMAT(I5,2F10.3)
103       FORMAT(3E20.8)
          READ(5,*)E,G,RHO
104       FORMAT(6E20.8)
105       FORMAT(2I10)
106       FORMAT(4I10)
107       FORMAT(E20.8)
          READ(5,*)CSAX,CSAY,SMAX,SMAY,TCX,TCY,POLX,POLY
          WRITE(6,132)E,G,RHO,CSAX,CSAY,SMAX,SMAY,TCX,TCY,POLX,POLY
132       FORMAT(///11X,25H***MATERIAL PROPERTIES***//
         1 5X,21HYOUNGS MODULUS* * * *,E16.8/
         2 5X,21HMODULUS OF RIGIDITY *,E16.8/
         3 5X,21HDENSITY * * * * * * *,E16.8/
         4 5X,21HC.S.A X-MEMBERS * * *,E16.8/
         5 5X,21HC.S.A Y-MEMBERS * * *,E16.8/
         6 5X,21HS.M.A X-MEMBERS * * *,E16.8/
         7 5X,21HS.M.A Y-MEMBERS * * *,E16.8/
         8 5X,21HT.CON X-MEMBERS * * *,E16.8/
         9 5X,21HT.CON Y-MEMBERS * * *,E16.8/
         8 5X,'POLAR 2ND MOM. OF AREA ABOUT X=',E16.8/
         7 5X,'POLAR 2ND MOM. OF AREA ABOUT Y=',E16.8)
          PI=3.141592654
          WRITE(6,133)
133       FORMAT(///4X,18H***ELEMENT DATA***//
         1 9X,1HI,9X,1HJ)
          DO 3 MEM=1,MEMS
          WRITE(*,99) MEM
99        FORMAT('   ELEMENT NO.',I5,,' UNDER COMPUTATION')
          READ(5,*)I,J
          WRITE(6,105)I,J
          MEMXP1=MEMX+1
          IF(MEM.GE.MEMXP1)GO TO 4
          A=CSAX
          SMA=SMAX
          TC=TCX
          POLAR=POLX
          GO TO 5
4         CONTINUE
          A=CSAY
          SMA=SMAY
          TC=TCY
          POLAR=POLY
5         CONTINUE
```

```
CALL NULL(SM,6,6)
CALL NULL(SK,6,6)
CALL NULL(SKX,6,6)
CALL NULL(DC,6,6)
CALL NULL(DCT,6,6)
XL=SQRT((X(J)-X(I))**2+(Y(J)-Y(I))**2)
DC(1,1)=1.0
DC(4,4)=1.0
DC(2,3)=(Y(J)-Y(I))/XL
DC(5,6)=DC(2,3)
DC(3,2)=-DC(2,3)
DC(6,5)=-DC(2,3)
DC(2,2)=(X(J)-X(I))/XL
DC(3,3)=DC(2,2)
DC(5,5)=DC(2,2)
DC(6,6)=DC(2,2)
SK(1,1)=12.0*E*SMA/(XL**3)
SK(4,4)=SK(1,1)
SK(4,1)=-SK(1,1)
SK(2,2)=G*TC/XL
SK(5,5)=SK(2,2)
SK(5,2)=-SK(2,2)
SK(3,3)=4.0*E*SMA/XL
SK(6,6)=SK(3,3)
SK(6,3)=SK(3,3)/2.0
SK(4,3)=6.0*E*SMA/XL**2
SK(6,4)=SK(4,3)
SK(3,1)=-SK(4,3)
SK(6,1)=-SK(4,3)
SM(1,1)=13.0/35.0+6.0*SMA/(5.0*A*XL**2)
SM(4,4)=SM(1,1)
SM(2,2)=POLAR/(3.0*A)
SM(5,5)=SM(2,2)
SM(3,1)=-11.0*XL/210.0-SMA/(10.0*A*XL)
SM(3,3)=XL**2/105.0+2.0*SMA/(15.0*A)
SM(6,6)=SM(3,3)
SM(4,1)=9.0/70.0-6.0*SMA/(5.0*A*XL**2)
SM(4,3)=-13.0*XL/420.0+SMA/(10.0*A*XL)
SM(5,2)=POLAR/(6.0*A)
SM(6,1)=13.0*XL/420.0-SMA/(10.0*A*XL)
SM(6,3)=XL**2/140.0-SMA/(30.0*A)
SM(6,4)=11.0*XL/210.0+SMA/(10.0*A*XL)
CN=RHO*A*XL
DO 6 II=1,6
DO 6 JJ=II,6
SK(II,JJ)=SK(JJ,II)
```

```
         SM(II,JJ)=SM(JJ,II)
6        CONTINUE
         CALL  SCPROD(SM,6,6,CN)
         CALL  MXTRAN(DCT,DC,6,6)
         CALL  MXPROD(SKX,SK,DC,6,6,6)
         CALL  MXPROD(SK,DCT,SKX,6,6,6)
         CALL  MXPROD(SKX,SM,DC,6,6,6)
         CALL  MXPROD(SM,DCT,SKX,6,6,6)
         I1=3*I-3
         J1=3*J-3
         DO 7 II=1,3
         DO 7 JJ=1,3
         MM=I1+II
         MN=I1+JJ
         NM=J1+II
         NN=J1+JJ
         STIFF(MM,MN)=STIFF(MM,MN)+SK(II,JJ)
         XMS(MM,MN)=XMS(MM,MN)+SM(II,JJ)
         STIFF(MM,NN)=STIFF(MM,NN)+SK(II,JJ+3)
         XMS(MM,NN)=XMS(MM,NN)+SM(II,JJ+3)
         STIFF(NM,MN)=STIFF(NM,MN)+SK(II+3,JJ)
         XMS(NM,MN)=XMS(NM,MN)+SM(II+3,JJ)
         STIFF(NM,NN)=STIFF(NM,NN)+SK(II+3,JJ+3)
         XMS(NM,NN)=XMS(NM,NN)+SM(II+3,JJ+3)
7        CONTINUE
3        CONTINUE
         WRITE(6,134)
134      FORMAT(///'DETAILS OF ZERO DISPLACEMENTS')
         DO 8 II=1,NFIXD
         READ(5,*)NPOSS(II),NXSUP(II),NYSUP(II),NZSUP(II)
         WRITE(6,106)NPOSS(II),NXSUP(II),NYSUP(II),NZSUP(II)
8        CONTINUE
         DO 777 II=1,NFIXD*3
         NSUPR(II)=0
777      CONTINUE
         DO 77 II=1,NFIXD
         IF(NXSUP(II).NE.0)NSUPR(II*3-2)=NPOSS(II)*3-2
         IF(NYSUP(II).NE.0)NSUPR(II*3-1)=NPOSS(II)*3-1
         IF(NZSUP(II).NE.0)NSUPR(II*3)=NPOSS(II)*3
77       CONTINUE
         READ(5,*)NCONC
         WRITE(6,677)NCONC
677      FORMAT('NO. OF ADDITIONAL CONCENTRATED MASSES=',I3)
         IF(NCONC.EQ.0)GO TO 678
         DO 679 II=1,NCONC
         READ(5,*)NPOS,XMCONC
```

```
      WRITE(6,701)NPOS,XMCONC
      I1=3*NPOS-2
      XMS(I1,I1)=XMS(I1,I1)+XMCONC
 679  CONTINUE
 701  FORMAT('NODAL POSITION OF MASS=',I3/
     1 'VALUE OF MASS=',E20.8)
 678  CONTINUE
      NFIXD3=NFIXD*3
      CALL SUPPRESS(STIFF,NSUPR,NJ3,NFIXD3)
      CALL INVMX(STIFF,NJ3,0.0,LVEC,MVEC,BVEC)
      CALL PRODAB(STIFF,XMS,NJ3)
      CALL NULL(XMS,NJ3,NJ3)
      D=0.001
      NP1=NJ3+1
      CALL EIGEN(STIFF,NJ3,XLAM,VEC,NP,D,XMS,NP1)
      NDF=3
      CALL EIGPRIN2(VEC,XLAM,NJ,NDF,NJ3,NP)
      RETURN
      END
C
C


C
C
      SUBROUTINE UNITMX(A,N)
      DIMENSION A(1)
C     (A)=(I)
      DO 1 I=1,N
      DO 1 J=1,N
      A((J-1)*N+I)=0.
      A((I-1)*N+I)=1.
  1   CONTINUE
      RETURN
      END
C
      SUBROUTINE NULL(A,M,N)
      DIMENSION A(M,N)
      DO 1 I=1,M
      DO 1 J=1,N
      A(I,J)=0.0
  1   CONTINUE
      RETURN
      END
C
      SUBROUTINE MXPROD(AB,A,B,N,M,L)
      DIMENSION AB(N,L),A(N,M),B(M,L)
```

```
C       (AB) = (A) * (B)
        DO 1 I=1,N
        DO 1 J=1,L
        AB(I,J)=0.0
   1    CONTINUE
        DO 2 I=1,N
        DO 2 K=1,L
        DO 2 J=1,M
        AB(I,K)=AB(I,K)+A(I,J)*B(J,K)
   2    CONTINUE
        RETURN
        END
C
        SUBROUTINE VALEIG(A,Y,X2,N,D,N1)
        DIMENSION A(1),Y(1),
       2Z(200),YO(200)
        X1=100000.0
        DO 1 I=1,N
        Y(I)=1.0
   1    CONTINUE
        X2=-100000.0
   2    CONTINUE
        DO 20 I=1,N
        YO(I)=Y(I)
  20    CONTINUE
        DO 4 I=1,N
        SUM=0.0
            DO  5  J=1,N
            SUM=SUM+A((J-1)*N1+I)*Y(J)
   5        CONTINUE
        Z(I)=SUM
   4    CONTINUE
        X2=0.0
        DO 21 I=1,N
        IF(ABS(X2).LT.ABS(Z(I))) X2=Z(I)
  21    CONTINUE
        DO 22 I=1,N
        Y(I)=Z(I)/X2
  22    CONTINUE
        IF(ABS((X1-X2)/X2).GT.D) GO TO   6
        GO TO 66
   6    CONTINUE
        X1=X2
          GO TO   2
  66    CONTINUE
        X = 0.0
```

```
      DO 11 I=1,N
      IF(ABS(X).LT.ABS(Y(I))) X=Y(I)
11    CONTINUE
      DO 12 I=1,N
      Y(I)=Y(I)/X
12    CONTINUE
      RETURN
      END
C
      SUBROUTINE EIGEN(A,N,AM,VEC,M1,D,XX,NP1)
      DIMENSION A(1),
     1 VEC(1),AM(1),XX(1),
     2 X(200),IG(200)
      MN=N
101   FORMAT(E20.8)
      NN = N
      CALL VALEIG(A,X,XM,NN,D,N)
      M=1
      DO 2 I=1,NN
      I3=(M-1)*N+I
      VEC(I3)=X(I)
      XX(I3)=VEC(I3)
2     CONTINUE
      AM(M) = XM
      IF(M1.LT.2) RETURN
      DO 20 M=2,M1
      DO 4 I=1,NN
      IF (ABS(XX((M-2)*N+I)-1.0).LT.0.0001)IR=I
4     CONTINUE
      IG(M-1)=IR
      DO 5 I=1,NN
      XX((MN-M+2)*N+MN-I+1)=A((I-1)*N+IR)
5     CONTINUE
      DO 6 I=1,NN
      DO 6 J=1,NN
      I3=(J-1)*N+I
      I4=(M-2)*N+I
      I5=(MN-M+2)*N+MN-J+1
      A(I3)=A(I3)-XX(I4)*XX(I5)
6     CONTINUE
      DO 9 I=1,NN
      IF(I.EQ.IR) GO TO 8
      IF(I.GT.IR) K1=I-1
      IF(I.LE.IR) K1=I
            DO 7 J=1,NN
            IF(J.EQ.IR) GO TO 7
```

```
                 IF(J.GT.IR) K2=J-1
          IF(J.LE.IR) K2=J
                 A((K2-1)*N+K1) = A((J-1)*N+I)
7                CONTINUE
8       CONTINUE
9        CONTINUE
        NN = NN-1
         M3=NN
         IF (M.NE.MN)  GO TO  11
         XM=A(1)
         X(1)=1.0
         GO TO  12
11      CALL VALEIG(A,X,XM,NN,D,N)
        WRITE(6,101)XM
12       DO 13 I=1,NN
         XX((M-1)*N+I)=X(I)
13       CONTINUE
         AM(M)=XM
         M11 = M-1
        MM1=1000-M11
        DO 20 M8=MM1,999
         M12 = M3+1
        M2=1000-M8
         M22 =IG(M2)+1
        MM22=1000-M22
        MM12=1000-M12
        IF(M12.LT.M22)GO TO 14
        DO 15 I3=MM12,MM22
        I=1000-I3
         X(I)=X(I-1)
15       CONTINUE
14       CONTINUE
        J=IG(M2)
        X(J)=0.0
        SUM=0.0
        DO 16 I= 1,M12
        SUM=SUM+XX((MN-M2+2-1)*N+MN-I+1)*X(I)
16       CONTINUE
        XK=(AM(M2)-XM)/SUM
        DO 17 I= 1,M12
        X(I)=XX((M2-1)*N+I)-XK*X(I)
17       CONTINUE
        SUM=0.0
        DO 18 I=1,M12
        IF (ABS(SUM).LT.ABS(X(I)))SUM=X(I)
18       CONTINUE
```

```
       DO 19 I=1,M12
       X(I)=X(I)/SUM
19     CONTINUE
       M3=M3+1
       IF (M2.NE.1)  GO TO 20
       DO 3 I=1,M3
       VEC((M-1)*N+I)=X(I)
3      CONTINUE
20     CONTINUE
       RETURN
       END
C
       SUBROUTINE INVMX(A,N,D,L,M,B)
       DIMENSION A(1),B(1),L(1),M(1)
       I5=0
3      I1=N-1
       DO 1 I=1,I1
       I2=I+1
       DO 2 J=I2,N
       I3=(I-1)*N+J
       I4=(J-1)*N+I
       B(J)=A(I3)
       A(I3)=A(I4)
2      A(I4)=B(J)
1      CONTINUE
       IF (I5.EQ.1) RETURN
       NK=-N
       DO 80 K=1,N
       NK=NK+N
       L(K)=K
       M(K)=K
       KK=NK+K
       BIGA=A(KK)
       DO 20 J=K,N
       IZ=N*(J-1)
       DO 20 I=K,N
       IJ=IZ+I
10     IF (ABS(BIGA)-ABS(A(IJ))) 15,20,20
15     BIGA=A(IJ)
       L(K)=I
       M(K)=J
20     CONTINUE
       J=L(K)
       IF (J-K) 35,35,25
25     KI=K-N
       DO 30 I=1,N
```

```
         KI=KI+N
         HOLD=-A(KI)
         JI=KI-K+J
         A(KI)=A(JI)
30       A(JI)=HOLD
35       I=M(K)
         IF (I-K) 45,45,38
38       JP=N*(I-1)
         DO 40 J=1,N
         JK=NK+J
         JI=JP+J
         HOLD=-A(JK)
         A(JK)=A(JI)
40       A(JI)=HOLD
45       IF (BIGA) 48,46,48
46       D=0.0
         RETURN
48       DO 55 I=1,N
         IF (I-K) 50,55,50
50       IK=NK+I
         A(IK)=A(IK)/(-BIGA)
55       CONTINUE
         DO 65 I=1,N
         IK=NK+I
         HOLD=A(IK)
         IJ=I-N
         DO 65 J=1,N
         IJ=IJ+N
         IF(I-K) 60,65,60
60       IF (J-K) 62,65,62
62       KJ=IJ-I+K
         A(IJ)=HOLD*A(KJ)+A(IJ)
65       CONTINUE
         KJ=K-N
         DO 75 J=1,N
         KJ=KJ+N
         IF (J-K) 70,75,70
70       A(KJ)=A(KJ)/BIGA
75       CONTINUE
         D=D*BIGA
         A(KK)=1.0/BIGA
80       CONTINUE
         K=N
100      K=(K-1)
         IF (K) 150,150,105
105      I=L(K)
```

```
         IF (I-K) 120,120,108
  108    JQ=N*(K-1)
         JR=N*(I-1)
         DO 110 J=1,N
         JK=JQ+J
         HOLD=A(JK)
         JI=JR+J
         A(JK)=-A(JI)
  110    A(JI)=HOLD
  120    J=M(K)
         IF (J-K) 100,100,125
  125    KI=K-N
         DO 130 I=1,N
         KI=KI+N
         HOLD=A(KI)
         JI=KI-K+J
         A(KI)=-A(JI)
  130    A(JI)=HOLD
         GO TO 100
  150    I5=1
         GO TO 3
         END
C
         SUBROUTINE PRODAB(A,B,N)
         DIMENSION A(N,N),B(N,N),VEC(500,1)
C        (A)=(A)*(B)
         DO 1 I=1,N
         DO 2 J=1,N
         VEC(J,1)=0.0
         DO 2 K=1,N
         VEC(J,1)=VEC(J,1)+A(I,K)*B(K,J)
  2      CONTINUE
         DO 1 J=1,N
         A(I,J)=VEC(J,1)
  1      CONTINUE
         RETURN
         END
C
         SUBROUTINE MXTRAN(AT,A,M,N)
         DIMENSION AT(N,M),A(M,N)
C        (AT)=(A)T
         DO 1 I=1,M
         DO 1 J=1,N
         AT(J,I)=A(I,J)
  1      CONTINUE
         RETURN
```

```
      END
C

      SUBROUTINE SCPROD(A,M,N,CN)
      DIMENSION A(1)
C     (A)=CN*(A)
      DO 1 I=1,M
      DO 1 J=1,N
      I3=(J-1)*M+I
      A(I3)=A(I3)*CN
    1 CONTINUE
      RETURN
      END
C

      SUBROUTINE MXNEG(A,B,M,N)
      DIMENSION A(M,N),B(M,N)
C     (A)=-(B)
      DO 1 I=1,M
      DO 1 J=1,N
      A(I,J)=-B(I,J)
    1 CONTINUE
      RETURN
      END
C

      SUBROUTINE MXSUM(A,B,C,M,N)
      DIMENSION A(M,N),B(M,N),C(M,N)
C     (A)=(B)+(C)
      DO 1 I=1,M
      DO 1 J=1,N
      A(I,J)=B(I,J)+C(I,J)
    1 CONTINUE
      RETURN
      END
C

      SUBROUTINE SUPPRESS(A,NSUP,N,NFIXD)
      DIMENSION A(N,N),NSUP(1)
      DO 10 II=1,NFIXD
      NS=NSUP(II)
      IF(NS.NE.0)A(NS,NS)=A(NS,NS)*1.0E12+1.0E12
   10 CONTINUE
      RETURN
      END
C
C
      SUBROUTINE  EIGPRIN2(VEC,XLAM,NJOINTS,NDOF,N,NV)
C
C     PRINTS RESULTS OF EIGENVALUE ANALYSIS:
```

```
C       FREQUENCY AND MODE SHAPE,INCORPORATING SUPPRESSED DOF'S.
C
        DIMENSION  VEC(N,NV),XLAM(NV)
        WRITE(6,30)
 30     FORMAT(///12X, 20H***EIGEN SOLUTION***)
        PI=3.1415927
        DO 150 IEIG=1,NV
          PAR=1.0/(2.0*PI)*SQRT(1.0/XLAM(IEIG))
        WRITE(6,60) IEIG, PAR
 60     FORMAT(///7X, I3, 1X, 12H FREQUENCY =,E14.6/
       1 8X, 29H----------------------------)
          WRITE(6,80)
 80        FORMAT(//10X,11H MODE SHAPE/
       1 10X,6H  NODE,5X, 37H RELATIVE DISPLACEMENTS(GLOBAL COORD))
          DO 90 IPOIN=1,NJOINTS
        IP=IPOIN*NDOF-NDOF
          WRITE(6,120) IPOIN, (VEC(IP+IDOF,IEIG),IDOF=1,NDOF)
 90        CONTINUE
 120    FORMAT(9X, I5, 3X, E10.3, 4X, E10.3, 4X, E10.3)
 150    CONTINUE
        RETURN
        END
```

A 6.2 A Typical Input File for GRIDIN.DAT (Example 9.2)

```
12   6   3
9   4
0.0   0.0
4.5   0.0
1.0   2.0
5.5   2.0
1.5   3.0
6.0   3.0
1.5   0.0
3.0   0.0
2.5   2.0
3.0   3.0
4.0   2.0
4.5   3.0
2E11   7.69E10   7860
4.0E-3   4.0E-3   1.25E-5   1.25E-5   2.5E-5   2.5E-5   2.5E-5   2.5E-5
2.5E-5
1   7
```

```
7    8
8    2
3    9
9   11
11   4
5   10
10  12
12   6
7    9
8   11
9   10
11  12
1    1   0   0
2    1   0   0
3    1   0   0
4    1   0   0
5    1   0   0
6    1   0   0
2
9   50
11  40
```

A 6.3 A Typical Output File for GRIDOUT.OUT (Example 9.2)

```
          ***BASIC PARAMETERS***

     NUMBER OF NODES * * * * * *12
     HNUMBER OF NODES WITH ZERO DISPLACEMENTS= 6
     NUMBER OF EIGEN VALUES* * * 3
     NUMBER OF X DIRECTION MEMS* 9
     NUMBER OF Y DIRECTION MEMS* 4



     ***NODAL POINT DATA***

     NODE      X          Y
        1    0.000      0.000
        2    4.500      0.000
        3    1.000      2.000
        4    5.500      2.000
        5    1.500      3.000
        6    6.000      3.000
        7    1.500      0.000
```

```
8      3.000      0.000
9      2.500      2.000
10     3.000      3.000
11     4.000      2.000
12     4.500      3.000
```

MATERIAL PROPERTIES

```
YOUNGS MODULUS* * * *   0.20000000E+12
MODULUS OF RIGIDITY *   0.76899999E+11
DENSITY * * * * * * *   0.78600001E+04
C.S.A X-MEMBERS * * *   0.40000000E-02
C.S.A Y-MEMBERS * * *   0.40000000E-02
S.M.A X-MEMBERS * * *   0.12500000E-04
S.M.A Y-MEMBERS * * *   0.12500000E-04
T.CON X-MEMBERS * * *   0.25000000E-04
T.CON Y-MEMBERS * * *   0.25000000E-04
POLAR 2ND MOM. OF AREA ABOUT X=  0.25000000E-04
POLAR 2ND MOM. OF AREA ABOUT Y=  0.25000000E-04
```

ELEMENT DATA

I	J
1	7
7	8
8	2
3	9
9	11
11	4
5	10
10	12
12	6
7	9
8	11
9	10
11	12

```
DETAILS OF ZERO DISPLACEMENTS
        1          1        0        0
        2          1        0        0
```

```
           3            1            0            0
           4            1            0            0
           5            1            0            0
           6            1            0            0
NO. OF ADDITIONAL CONCENTRATED MASSES=   2
NODAL POSITION OF MASS=  9
VALUE OF MASS=          0.50000000E+02
NODAL POSITION OF MASS= 11
VALUE OF MASS=          0.40000000E+02
         0.63945454E-04
         0.10408248E-04
```

```
              ***EIGEN SOLUTION***



      1   FREQUENCY =   0.154642E+02
      -------------------------------



           MODE SHAPE
            NODE       RELATIVE  DISPLACEMENTS(GLOBAL COORD)
             1     0.857E-13       0.142E-01     -0.700E+00
             2     0.113E-12       0.233E+00      0.768E+00
             3     0.115E-12      -0.227E+00     -0.811E+00
             4     0.118E-12       0.479E-01      0.822E+00
             5     0.108E-12      -0.261E+00     -0.733E+00
             6     0.749E-13      -0.536E-01      0.647E+00
             7     0.881E+00       0.142E-01     -0.371E+00
             8     0.928E+00       0.232E+00      0.329E+00
             9     0.990E+00      -0.227E+00     -0.366E+00
            10     0.886E+00      -0.260E+00     -0.315E+00
            11     0.100E+01       0.479E-01      0.366E+00
            12     0.823E+00      -0.536E-01      0.360E+00



      2   FREQUENCY =   0.199029E+02
      -------------------------------



           MODE SHAPE
            NODE       RELATIVE  DISPLACEMENTS(GLOBAL COORD)
             1     0.123E-12      -0.635E+00     -0.815E+00
```

2	0.131E-12	-0.542E+00	0.838E+00
3	-0.962E-14	-0.600E+00	0.137E+00
4	-0.414E-13	-0.615E+00	-0.222E+00
5	-0.115E-12	-0.537E+00	0.703E+00
6	-0.958E-13	-0.619E+00	-0.650E+00
7	0.982E+00	-0.634E+00	-0.351E+00
8	0.100E+01	-0.541E+00	0.340E+00
9	-0.187E+00	-0.599E+00	0.103E+00
10	-0.829E+00	-0.536E+00	0.264E+00
11	-0.253E+00	-0.614E+00	-0.646E-01
12	-0.788E+00	-0.618E+00	-0.288E+00

3 FREQUENCY = 0.493323E+02

MODE SHAPE

NODE	RELATIVE	DISPLACEMENTS	(GLOBAL COORD)
1	0.923E-13	-0.665E+00	-0.361E+00
2	0.269E-13	-0.605E+00	0.238E+00
3	-0.803E-13	0.420E+00	0.304E+00
4	-0.570E-13	0.476E+00	-0.249E+00
5	0.724E-13	0.941E+00	-0.419E+00
6	0.123E-12	0.100E+01	0.512E+00
7	0.378E+00	-0.658E+00	-0.701E-01
8	0.309E+00	-0.598E+00	0.171E+00
9	-0.312E+00	0.416E+00	0.467E-01
10	0.500E+00	0.931E+00	-0.210E+00
11	-0.273E+00	0.470E+00	-0.736E-01
12	0.548E+00	0.989E+00	0.126E+00

References

1. Turner MJ, Clough RW, Martin HC, Topp LJ (1956) Stiffness and deflection analysis of complex structures. J Aero Sci 23:805–823
2. Sack RL (1989) Matrix structural analysis. PWS-Kent Publishing Co, Boston
3. Rockey KC, Evans HR, Griffiths DW, Nethercot DA (1985) The finite element method, 2nd edn. Collins, Glasgow
4. Zienkiewicz OC, Taylor RL (1989) The finite element method. McGraw-Hill, New York
5. Courant R (1943) Variational methods for the solution of problems of equilibrium and vibration. Bull Am Math Soc 49:1–23
6. Ross CTF (1987) Applied stress analysis. Horwood
7. Ross CTF (1990) The finite element method in engineering science. Horwood, Chichester
8. Ross CTF (1977–1978) Algol programs for structural vibrations. TASS, AUEW, Richmond, Surrey
9. Irons B (1965) Structural eigenvalue problems: elimination of unwanted variables. JAIAA 3:961
10. Ross CTF (1987) Advanced applied stress analysis. Horwood, Chichester
11. Clarkson J (1959) Data sheets for the elastic design of flat grillages under uniform pressure. European Ship-building, no. 8
12. Venancio-Filho F, Iguti F (1973) Vibrations of grids by the finite element method. Computers and Structures 3:1331–1344

Index